Fireman's World

Anne Burridge

MERLIN BOOKS LTD.
Braunton   Devon

I dedicate this book to our firemen . . . .
For all they have to know
And all they have to be;
For all they have to do
And all they have to see.

Anne Burridge, 1981

ISBN 0 86303 067-X
Printed in England by Maslands Ltd., Tiverton, Devon

CONTENTS

Colour section set between pp 56 & 57

ACKNOWLEDGEMENTS

The boundless co-operation I've received while compiling this book, both from Fire Service personnel and others, makes it impossible to mention everyone involved. I can only thank from the heart all who have played a part in it, however small.

Most of my research has been done in Bucks, and I am especially indebted to the following personnel of Buckinghamshire Fire Brigade: County Fire Officer A. J. Archer, Q.F.S.M., F.I. Fire E., for granting me unlimited facilities; Divisional Commander Derek Dodwell, G.I. Fire E., for putting himself at my disposal for advice and assistance throughout; the Station Officers and personnel of High Wycombe Fire Station and Beaconsfield Fire Station for their wonderful co-operation, and also to the control staff at Aylesbury Control Room; Station Officer John Maher of the Fire Prevention Department for his vast amount of help in advising me and taking photographs, and to L/Fm. Martin Mawhood and L/Fm. Tony Cairns also for taking photographs. These three amateur photographers very kindly gave up their time to take the large number of photographs that appear without credits. My thanks too, to Assistant Divisional Officer James Henderson for photographic material.

I owe a special debt to Fm. F. G. E. Stanley (Beaconsfield) and his wife who kept open house week after week to help me with my first research.

Many others outside Bucks have given invaluable assistance, notably the Fire Service Inspectorate at the Home Office. I am particularly grateful to H.M. Inspector of Fire Services Robert Clark, M.M., M.I. Fire E., and H.M. Assistant Inspector of Fire Services Brian Webb, F.I. Fire E., for their unstinting co-operation.

I am also indebted to the following: the Commandant of the Fire Service College, Mr David Blacktop, C.B.E., F.I. Fire E., for permitting photographs; *Fire* Magazine for their most willing assistance and for all I have learned through their pages — in particular for use of details from their Supplement on the Moorgate disaster; Assistant Chief Fire Officer Ivan Henson of Humberside Fire Brigade for use of his Report of the Flixborough explosion; Mr J. H. Beckers, General Manager, Nypro (U.K.) Ltd., for consenting to use of the Site Plan; to Temp. L/Fm. Chris Nelson (F.S. Training Centre, Maidstone) and Mr Owen Rowland (King's College, London University) for photographs used gratis; and to Mr Eric George and Fm. Jaime Graham (Kentish Town F/Stn.) for photographing drills.

The following have also given invaluable help: Station Officer Braxton and Temp. Station Officer John Barker with the men of the Red Watch, Paddington Fire Station; Hazchem Section, London Fire Brigade; L/Fm. Newell, London Fire Brigade Photographic Library; the British Standards Institution; H.M. Stationery Office; Station Officer Paul Taylor (Hayes Fire Station); all personnel discussed in the chapter on injuries; all brigades and other parties represented by photographs, and in particular the brigades of Essex, Hampshire, London, and Tyne and Wear whose involvement, after that of Buckinghamshire, has been exceptional.

Finally, special thanks to my husband and family for their support.

Anne Burridge

INTRODUCTION

It's probably true to say that no one has a more dangerous job than the fireman. It must certainly be true that no one has a more varied job. For a fireman's work may take him from burning factory to forest fire; from chemical spill to flood. He may find himself rescuing victims of air and rail disasters; people trapped in factory machinery; others in wrecked cars. He'll be needed when someone gets stranded in a lift or when a frightened youngster gets his head stuck between railings.

Where there are people there are accidents, and they happen in a hundred different ways, so a fireman must be able to turn his hand to anything. Often, in life or death situations, he's the only one left to turn to for help, and he must know exactly what to do. With his assortment of special equipment and the know-how that goes with it, he must work against impossible odds and win.

His courage and endurance, his compassion for others and lack of thought for himself, set him apart as a very special man. He goes about his work without fuss and often without thanks, yet his value to the rest of us is beyond measure.

My book is an attempt to show you something of the fireman's world and all he does for us, often at risk to his own safety.

A LOOK ROUND THE FIRE STATION

Up and down the country are fire stations like this one. They never close. The men who man them are on call round the clock, and they'll turn out in the dead of night or in the middle of dinner. Whether it be Christmas Day or Boxing Day, Eastertime or Whitsun Holiday, whenever the rest of us are enjoying ourselves, they'll be there. We send for them — they come running. We ask the impossible of them — they never fail us. Their work is dirty, exhausting and dangerous. Yet ask any fireman about his work and he'll tell you he enjoys it. This is partly because the job varies from day to day and firemen never know what they'll be asked to do next! Besides, they enjoy the companionship of teamwork. Most of all, they see their job as an opportunity to make themselves useful to the rest of us. Lucky for us! Where would we be without them?

Like any apprentice learning a special skill, a fireman has to study hard. He reads and attends lectures as well as training with his equipment. This training continues even after he's passed all his examinations and is a qualified fireman, for he must move with the times. Equipment he uses today may be out of date in a year's time; new building materials are being introduced all the time and this may affect how he fights certain fires; also, the number of dangerous chemicals he'll have to deal with increases year by year — and there are already over 15,000 of these!

Later on, you'll see photographs of firemen training and working at all kinds of incidents, but since you've probably never taken a really good look inside a fire station, let's begin there and do some exploring.

You'd be surprised how many rooms there are in a fire station! Locker rooms and cloakrooms; store rooms and repair rooms; wash rooms; offices; rooms to study in, eat in, relax in. And of course, accommodation for the most important residents of all — the fire engines (or, as the fireman calls them, the appliances). We are going to take a closer look at those rooms which play the largest part in the day to day running of the fire station:

1 — the Appliance Room
2 — the Watch Room
3 — the Lecture Room
4 — the Mess Room
5 — the Kitchen
6 — the Drying Room
7 — the Dormitory
8 — the Pole House
9 — the Compressor Room, and
10 — the Hose Repair Room.

A Water Tender stands ready in the appliance room.

The Watch Room, where a fireman is taking down a message at the switchboard. Note the numbered maps in the rack by the door. Others are pinned on the walls.

THE APPLIANCE ROOM

This where the action starts! The appliances are kept ready for a fast turnout with cab doors left open and the firemen's boots standing alongside with leggings rolled down over them ready for the men to step into.

THE WATCH ROOM

Every fire station has a Watch Room. This is the communications centre linking the fire station with its local headquarters and (through headquarters) with other fire stations in the area. In the Watch Room are numerous detailed maps of the area served by the fire station. Firemen consult these regularly to pinpoint various addresses. In here, too, the firemen's driving hours are logged. In fact the Watch Room is really a kind of firemen's office.

THE LECTURE ROOM

Though not every fire station has a Lecture Room as such, a suitable room, such as the Quiet Room, is usually set aside for this.

THE MESS ROOM

At all fire stations this acts as the general meeting place and is also where firemen have their meals.

THE KITCHEN

No fire station would be complete without a kitchen! Firemen club together to pay for their meals, then elect one of their number to buy and cook the food. He's known officially as the Mess Manager and holds the post as long as he is willing and his cooking is tolerable! His is one of the less popular jobs around the fire station, not only because his colleagues

Hungry colleagues make short work of a meal . . .

. . . prepared by the Mess Manager.

A fire station's dormitory, showing fold-away beds.

The quickest way down — a fireman slides down the pole. The protective cushion around its base ensures a soft landing!

A compressor, used to fill cylinders with compressed air for use in breathing apparatus.

have critical palates, but also because he gets left behind peeling potatoes when he'd rather be where the action is.

DRYING ROOM
A fireman's uniform is often soaked through when he returns from a fire, so there's a special drying room where he can off-load his wet clothing. Firemen are issued with two spare uniforms.

DORMITORY
The night shift is 15 hours long (6 p.m. – 9 a.m.) so a dormitory is provided where firemen can rest fully clothed with boots off, in readiness for incoming calls. Quite often they're out all night dealing with fires and road accidents. Calls in the middle of the night are rarely false alarms, and the fact that fire often remains undiscovered for some time while people are asleep means that the incidents tend to be bigger.

POLE HOUSE
Leading straight down from the upper floors to the appliance room on the ground floor is a steel pole, around which at all floor levels is an opening large enough for firemen to climb on to it and slide down. (I've often wanted to have a go on this!) It's faster and safer than hurrying down the stairs, especially at night. The access area to the pole is officially called the POLE HOUSE. If you're a fireman you call it the 'pole hole'. And by the way, contrary to what people think, the pole is not polished!

THE COMPRESSOR ROOM
Some fire stations have a compressor for charging breathing apparatus cylinders with compressed air. The compressor is housed in a room on its own, and air is fed through pipes to the machine on the other side of the wall. Connectors on the feed pipes are detached from their holders and screwed on to the necks of the cylinders to be filled. The valves are turned on, and the compressor is started up. Several cylinders can be charged at once. Those that have some air left in them will fill first and air will overflow into the others until eventually all of them are full. When they are at the correct pressure (3,000 lbs) the compressor switches off automatically.

THE HOSE REPAIR ROOM
From time to time hose becomes chafed or otherwise damaged during its working life and eventually may burst. The hose then becomes less efficient for firefighting as water pressure is lost, so it goes to the hose room to be repaired.

A special electrically heated machine is used for this. The hose is laid in the machine with the patch uppermost, and the heated pad is then pressed down onto it, fusing the materials together. The heating element is thermostatically controlled so that just enough heat is applied for each individual repair job. Learning how to repair hose is all part of a fireman's work.

APPLIANCES

Firemen are renowned for the smartness of their appliances. Not only must they look good — the equipment in the lockers has to be checked regularly in case any tools or materials need replacing. A fire appliance is classed as a heavy goods vehicle and in keeping with the rules of the road, only firemen over the age of 21 are allowed to drive them.

Here are some of the appliances firemen use.
1 — Water Tender
2 — Pump Escape
3 — Turntable Ladders
4 — Hydraulic Platform
5 — Emergency Tender
6 — Control Unit.

WATER TENDER
This appliance carries between 1800/2250 litres of water and hosereel equipment to tackle the fire straight away. It has its own built-in pump but carries a portable pump as well. Its lightweight aluminium extension ladders can reach a height of 10.5 or 13.5 metres. A wide variety of tools is stowed in the lockers.

PUMP ESCAPE
Like the Water Tender, this appliance has its own supply of water and hosereel equipment to attack the fire on arrival, but carries less water (a minimum of 455 litres) because of the weight of its wooden ladders. The wheeled escape, weighing 762 kg is removed from the appliance and positioned as necessary. The ladders, which are extended by steel cables, reach a height of 13.7 – 15 metres. These appliances are becoming rarer, partly because the large wheels are a problem where access is limited, and partly because of the difficulty of maintaining the wheels themselves.

TURNTABLE LADDER
This is a sectional ladder extending to 30.5 metres and mounted on a platform that can rotate through a complete circle. It is operated mechanically or hydraulically and requires a crew of two or more men, one of whom is specially trained to operate the controls. It is equipped with

Two breathing apparatus cylinders being filled with compressed air by the compressor in the next-door room. 8 cylinders can be charged on this machine. The large nozzles at the end of the black feed pipes are disconnected from the blue holders and connected to the necks of the cylinders. The red taps operate the valves allowing air to flow into them.

A hose repair room, showing hose cooling on the floor after being repaired with special patches using the machine on the left. The hose is laid on the lower pad with the patch in position, then the upper pad, heated like an iron, is pressed onto it, sealing the patch by heat.

A Water Tender

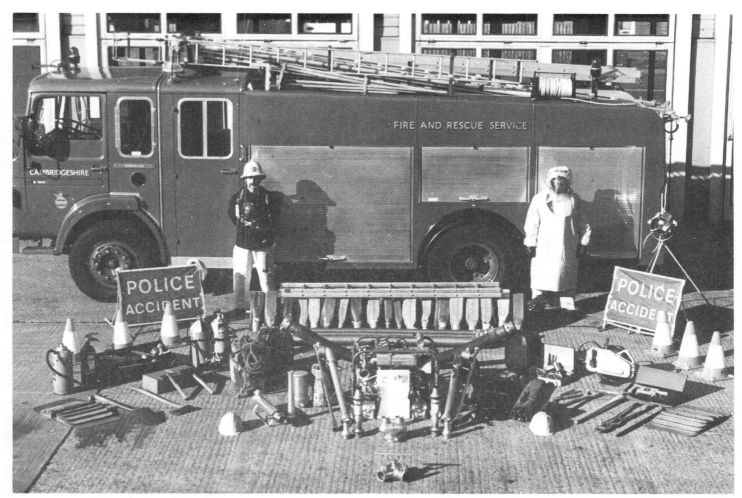

The range of equipment typically carried by a Water Tender, as displayed by Cambridgeshire Fire and Rescue Service.

Photo: Cambridgeshire Fire and Rescue Service

A Pump Escape

Here, the escape ladder has been taken off the machine and pitched to the drill tower.

The Turntable Ladders fully extended.

a MONITOR that can direct a continuous deluge from the head of the ladder on to the fire, acting as a 'water tower'. The Turntable Ladder is also used for rescue work. Apart from bringing trapped people down the ladder itself, firemen can use the 76.25 metre rescue line carried on the appliance to haul people to safety, by passing it through a special fitting at the head of the ladder.

In high winds the great height of the ladder makes it very dangerous to work on, and the head must be stabilized by at least two guy lines held by firemen. Jacks provide stability while the ladder remains extended.

HYDRAULIC PLATFORM
This appliance serves basically the same purposes as the Turntable Ladder. It has hydraulically operated booms, the upper

The Turntable Ladders

The Hydraulic Platform, or Snorkel.

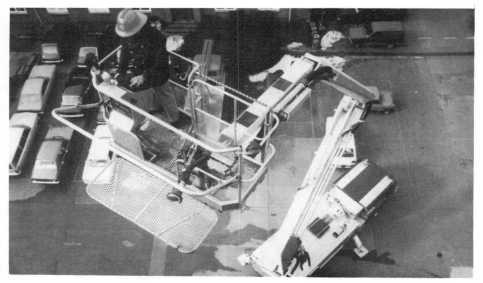

Northamptonshire's Simon Snorkel SS220 during an exercise. The fireman in the cage is using controls at the console to operate the booms. Below, a second fireman is maintaining contact with the cage. Photo: Simon Engineering, Dudley Ltd.

The Emergency Tender with its distinctive chequered back.

one being fitted with a cage from which the monitor can deliver 4,546 litres of water per minute. Capable of rotating through a complete circle like the Turntable Ladder, the booms can be controlled either from the cage or from the ground. In an emergency the ground controls are used to override those in the cage. For rescue purposes the cage accommodates about five people. This appliance comes in various sizes, the most commonly used reaching a height of 19.8 – 22.9 metres and 24.33 – 26 metres respectively.

EMERGENCY TENDER

This the Jack-of-all-trades among the appliances, with special equipment aboard to deal with every kind of emergency. Its two major roles are at road accidents and chemical incidents. It also serves as a Control Unit at large fires and other major incidents.

Its equipment includes lighting facilities for use at night or when visibility is inadequate. It carries all kinds of cutting gear, including propane hot flame cutting equipment to tackle heavy machinery or wreckage in which people are trapped. Other rescue equipment includes jacks and inflatable air bags, winches, cables and ropes, as well as resuscitation apparatus and essential medical supplies.

For chemical incidents there are protective suits, Breathing Apparatus, pumps, foam-making equipment and so on. Salvage sheets are carried too, to protect factory machinery or storm-damaged homes. A large collection of hand tools is included for good measure.

CONTROL UNIT

A Control Unit acts as a mobile Headquarters at large incidents (5 pumps or over) and carries an assortment of radio equipment so that contact can be established both with the Brigade Control Room and with personnel at the incident. In addition to the normal VHF radio, it may carry walkie-talkies and field telephones.

A plan of the incident is set up in the Control Unit, and the Duty Officer and personnel assisting him keep account of the whereabouts of men and appliances and report on progress.

Nominal Roll Boards from individual fire stations recording the names of personnel involved are handed in together with the men's tallies, for collection when their duties are completed.

Some Control Units, like the one in the photograph, are purpose-built, but a Control Unit can be set up in any appliance equipped with the right facilities, though 'he Emergency Tender is the one most

An Emergency Tender belonging to Cambridgeshire Fire and Rescue Service, showing the extensive amount of equipment carried to meet every emergency.
Photo: Cambridgeshire Fire and Rescue Service

often used. A flashing red beacon atop the rear of the Control Unit enables it to be identified among the other appliances at the incident.

A Control Unit, showing its radio masts folded down on the roof.
Photo: Owen Rowland

The appliances we have been looking at form the backbone of our firefighting resources although I must stress that they are only a representative selection. You will also see numerous others, including hose-laying lorries, Chemical Incident Units and, if you live near a busy port, powerful fireboats that deal with fires in wharves and other riverside properties.

Fire gear.

Undress uniform.

UNIFORM

A fireman has two kinds of uniform: his FIRE GEAR, and his UNDRESS UNIFORM.

He wears his fire gear during drills and while attending incidents. The undress uniform is worn about the fire station and elsewhere while he's carrying out normal duties such as fire prevention work, visiting factories and so on.

Fire gear consists of a double-breasted fire tunic with silver buttons, made from thick navy-blue Melton cloth, a yellow helmet of fibreglass or cork with chinstrap (white for Station Officers and other senior ranks), yellow PVC leggings or overtrousers, and black rubber boots. He also wears a black silk scarf which, as well as preventing chafing from his tunic,

protects his neck against hot debris.

Undress uniform comprises a navy-blue suit with silver buttons, matching cap, blue shirt (white for Station Officers and above) and tie, socks and shoes all in black.

At incidents and at drills, fire gear is always worn over undress uniform — minus the cap, jacket and shoes of course! You might think, with all this cumbersome clothing on that a fireman is too well protected to get burned, but he's not made of asbestos and nor are his clothes! In a fire he's as vulnerable as you or I.

Rank markings for his uniform are shown on the next page.

Undress uniform — Silver
Fire uniform — Chrome

On lapel of jacket a gorget patch of black with a centre cord of red embroidered oak leaves

On lapel of jacket a gorget patch of black with a centre red cord

Chief Officer

Deputy Chief Officer

Chief Staff Officer and Assistant Chief Officer

Deputy Assistant Chief Officer

Senior Divisional Officer

Divisional Officer

Assistant Divisional Officer

Station Officer

Sub Officer

Leading Fireman

CONTROL STAFF

Principal Controller

Senior Controller

Area Controller

Control Officer I

Control Officer II

No shoulder markings for fireman rank

No shoulder markings for Control Officer III
All Control Staff markings on a red background

White helmet with 1½" black band

White helmet with two ¾" black bands with ½" separation

White helmet with ¾" & ½" black bands with ½" separation

White helmet with ¾" black band

Chief Officer

Deputy Chief Officer & Assistant Chief Officer

Deputy Assistant Chief Officer & Senior Divisional Officer & Divisional Officer

Assistant Divisional Officer

White helmet with ½" black band

Yellow helmet with two ½" black bands with ½" separation

Yellow helmet with one ½" black band

Yellow helmet

Station Officer

Sub Officer

Leading Fireman

Fireman

Caps	Chief Officer Two rows of silver oak leaves on peak	Assistant Divisional Officer Plain cloth peak
	Deputy & Assistant Chief Officer One row of silver oak leaves on peak	Station Officer Plain cloth peak
		Sub Officer Standard pattern fireman's cap
	Deputy Assistant Chief Officer and Divisional Officer ½" raised silver embroidery on peak	Leading Fireman Standard pattern fireman's cap

BREATHING APPARATUS

Breathing apparatus is as vital to a fireman as a space suit is to an astronaut.

It's his life support system when he's surrounded by fire and the air is too hot to breathe, as well as containing poisonous fumes and smoke. It's his protection too, when he's threatened by leaking gases and deadly chemicals.

It's a sophisticated piece of equipment with a host of individual parts, but for the sake of simplicity we shall look at only the most basic ones. Not all breathing apparatus sets *look* alike but they work in much the same way.

The cylinder (1) strapped to the fireman's back holds compressed air, fed along a tube into his face mask (2). Air passes to him through the demand valve (3) as he needs it, and the exhale valve (4) allows the used air to escape into the outside atmosphere.

The pressure gauge (5) keeps account of how much compressed air is left in his cylinder. The cylinder should be not less than 6/7 full at the start of operations. His air supply lasts him about 35 minutes, plus a 10-minute safety margin. When the 35 minutes are up, the warning whistle (6) goes off, sounding like a whistling kettle. The fireman must check his pressure gauge regularly and make sure he's back at the exit point by the time the whistle goes. If he fails to appear at the time he's due, it is hoped that the 10-minute safety margin will provide enough time for him to be rescued before his cylinder empties completely. If he runs into trouble and becomes trapped or injured, he can press a button on the Distress Signal Unit (7) — DSU for short — which sends out a loud, penetrating note to alert other firemen and guide them to him. Only when he's safely out will the B.A. Control Officer turn the signal off by unlocking the mechanism of the DSU with its matching key.

A Distress Signal Unit (DSU) showing tally to be filled in with details. The push button is on the underside.

A fireman's Breathing Apparatus showing its basic parts. (Front view).

Breathing Apparatus from the rear.

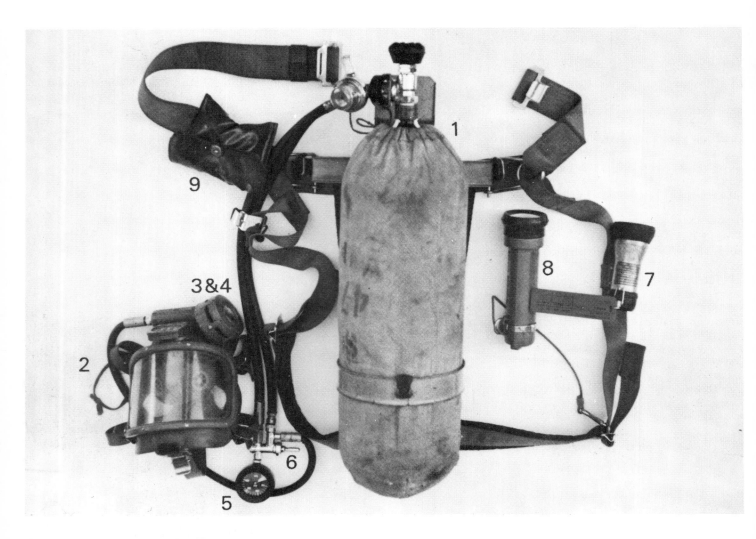

Typical breathing apparatus used by Firemen.

1. Cylinder containing compressed air.
2. Face mask.
3. and 4. Combined demand/exhale valves.
5. Pressure guage.

6. Low air pressure warning whistle.
7. Distress Signal Unit.
8. Safety torch.
9. Pouch containing personal guide line.

THE STATION OFFICER

The overall running of the fire station is the responsibility of the Sation Officer.

He's responsible for everyone on the premises, from his next-in-line the Sub Officers, down to Leading Firemen, Firemen and civilian staff such as part-time cooks and cleaners.

Although he gives authority to his senior men to see that the work of the fire station is carried out, the safety and efficiency of his men and appliances are his concern. Brigade Orders are drawn up by his county's Chief Fire Officer in conjunction with the Home Office in London and local authorities, and it's his job to see that they are carried out. These orders take into account any new laws that fire brigades must observe, and also ensure that firemen cover the curriculum as regards various drills and training programmes.

As Station Officer, one of his duties is to train other officers, which he does by encouraging his Sub Officers and Leading Firemen to take over from him in conducting drills, taking lectures and doing routine office work. While he's away his Sub Officers run the fire station, with his Leading Firemen assuming responsibility for keeping the appliances clean, fully equipped and topped up with fuel, oil and water. In fact they make sure that the routine work of the fire station is kept up to scratch and that things run smoothly.

Outside the fire station, the Station Officer directs his crews at local emergencies. It also falls to him to make fire safety inspections in the neighbourhood.

Back in his office, most of his time is spent reading and signing various forms and fire reports, authorizing supplies and dealing with brigade correspondence.

However busy he is, the Station Officer is always prepared to consider requests from the public to look around the fire station, and to send a courteous reply.

(In certain cities, because of the high life risk, fire stations have 4 Station Officers, one in charge of each Watch – Red, White Blue and Green. In this case each Watch has specific duties to carry out. But if an emergency arises and one Watch is unable to finish the work, the next Watch deals with it.)

The Station Officer always has plenty of paperwork to do.

ROTA AND DUTIES

Fire stations fall into three categories: 1) The full-time or shift fire station catering for busy areas such as large towns, and which is manned round the clock. 2) The day-manned fire station serving smaller towns, which is manned only throughout the day but giving part time cover at night as the need arises. 3) The part-time fire station found in more rural areas, and which is manned by part-time firemen who report as necessary.

Firemen at most full-time fire stations are grouped into Watches – Red, White, Blue and Green – and each Watch works an 8-day rota; two day shifts, two night shifts, and the next four days off. There being 7 days to a week, this 8-day pattern will start one day later each week, and it will take 8 weeks for the original rota to come round again.

Here, firemen work in two shifts, from 9 a.m. – 6 p.m. and from 6 p.m. – 9 a.m., unlike firemen at a day-manned fire station who are on duty from 9 a.m. – 5 p.m. and have a special rota to provide fire cover at all other times.

The work of the full-time men is backed up by a number of part-time firemen, who have regular jobs outside the Fire Service and are called out in special circumstances. For example, when two appliances have gone out to deal with a house fire and another fire arises, the part-time firemen will be called in by a signal over their personal pocket ALERTERS.

Again, if a large building such as a hospital or college is involved, and the required attendance is three machines, the part-time firemen will be called in to man the third machine. Part-timers operate under two systems – full cover, which means they can be available at all times, and three-quarter cover, where a part-timer may be employed on a factory assembly line and can only be available between certain times.

HOW A FIREMAN'S DAY IS SPENT

Since every Chief Officer has his own ideas about how his brigade should function and what equipment it should have, there are slight variations between the brigades. Bearing this in mind let's look at a typical timetable for a fireman on day shift.

TIMETABLE
9 – 9.30 a.m. – Roll Call and testing of equipment
9.30 – 10.45 a.m. – Drill

10.45 – 11 a.m. – Stand Easy (a 15 minute break for tea)
11 a.m. – 12 noon – Technical Training
12 noon – 1 p.m. – Lecture on Fire Service matters
1 p.m. – 2 p.m. – Midday meal
2 p.m. – 3.45 p.m. – One crew takes an appliance out to deal
with 'OFF STATION' duties. Remaining crews deal with 'ON
STATION' duties.
3.45 – 4 p.m. – Stand Easy for all crews (afternoon tea break)
4 p.m. – 6 p.m. – All uncompleted business is dealt with
6 p.m. – Change of shift
At any time of course, this well-ordered routine is likely to be
upset by an emergency, and this includes meals as well as
training! But the experience the fireman gains from going to
actual incidents is more valuable than the toughest training.
Let's look at this programme a bit at a time, beginning with
Roll Call.

ROLL CALL

At the start of each shift a special Test Call is put out to check
that the system is operating properly, after which firemen
coming on duty report to the appliance room fully uniformed
and stand at attention to answer their names as they are called.

After making sure that enough men are available to man the
appliances, the outgoing Watch is dismissed. In the event of
there being absentees due to illness or poor travelling
conditions, the appropriate number of men from the off-duty
watch must remain on duty until replacements can be found. At
this point, each on-duty fireman is told which appliance he's
riding for the day, for changes are constantly made with the
object of making all firemen familiar with the various
appliances and the work that goes with them.

TESTING OF EQUIPMENT

Following Roll Call all equipment is checked. The fire
appliances are started up and given a quick inspection to see
that the main water tanks are full, and that they are topped up
with fuel, oil and water. Equipment in the lockers must also
be checked to make sure that all essential items are available and
in working order. It's not unusual for small items to be mislaid
in the darkness at night-time incidents, and during the check-up
period, losses of this sort will come to light, so that they can be
made good. Breathing apparatus also has to be inspected
thoroughly to verify that there's no leakage of air from the
tubes, and that the equipment is generally in good order.
Torches are tested in readiness for searching in darkness or
smoke.

Much of a fireman's time is spent in training. This takes the
form of a) drills using the fire appliances, b) lectures on subjects
related to his work, and c) technical drills involving the use of
tools and special rescue equipment. First the drill.

DRILLS

A drill doesn't always follow the same pattern. Firemen are
constantly faced with different situations and have to know
how to handle each one as it arises. For example, sometimes
dealing with a fire means rescuing people, possibly from upstairs
rooms. At other times special equipment is called for. Drills
are accordingly varied to acquaint firemen with different
procedures and train them in the use of all kinds of equipment.
There are, in fact, over 40 different pre-planned drills laid down
in the curriculum, any one of which may be used, modified to
represent a particular situation. At Roll Call each fireman is
told his position on a certain appliance for the day: 1, 2, 3, 4
(and on some machines, 5). His duties are determined by this
position. So every fireman knows what he is expected to do and

The drill tower acts as the building.

This Leading Fireman will conduct the Drill.

20

can work together with the rest of the team so that the whole operation runs smoothly.

The procedure will be easier to follow if we watch firemen first working mainly with ladders, and then mainly with hose, though of course both duties are carried out simultaneously.

LADDERS

Without ladders, much of the fireman's work simply could not be done, since they allow him an independent means of access to points where this would otherwise be impossible, e.g. to the upper floors of buildings, both from outside where no stairs exist, and from inside where they have been destroyed by fire.

Sometimes he is called upon to make rescues from high structures such as bridges or towers, or possibly to rescue an animal from a tree or some other inaccessible place.

And of course, in the same way that ladders are useful for gaining access *upwards*, so they are useful for gaining access *downwards*, as will be seen from the two photographs in the section on general rescue work, the first showing the train crash at Invergowrie, where ladders are being used down the embank-ment; and the second showing firemen involved in the rescue of 11 workmen trapped underground, where ladders were used to bring the men to the surface.

Thus, a fireman's equipment includes a variety of ladders, some of which he manoeuvres by hand, and others which form part of the special appliance, the Turntable Ladders, which are used when extra height is needed.

Extension ladders, whose sections lie neatly one over the other while stowed on the fire appliances, can be extended by pulling downwards on the attached line (see following drills). These ladders must always be footed by another fireman while a member of the crew is working on them. Wherever possible, ladders are pitched to the right of the window, allowing a clear space to the left for rescue purposes. When moving up or down a ladder, a fireman uses left hand and foot in unison, then right hand and foot.

Ladders can be used for other purposes too. Sections of extension ladders can serve for bridging purposes — either singly or lashed together — providing access across rivers and streams, or awkward gaps (see page 24). Short sections of these ladders can also be used to form stretchers.

The   following   Drill   shows   ladders   in   use.

A LADDER DRILL (Using a 10.5m extension ladder — Sequence, 6 stages).

1. The fastenings are released and the ladder is slipped off the appliance. Rollers help it to run smoothly.

2. *The ladder is lifted clear and carried to the building.*

3. *A fireman stands on the bottom round of the ladder to hold it down as it is elevated.*

4. *He extends the ladder by pulling on the line as the others steady it.*

5. The pawls are engaged to secure the ladder in position.

6. The ladder is pitched to the window at a safe angle.

A fireman moves up the ladder using right hand and right foot, left hand and left foot. While he is working on it the ladder must be 'footed' by another fireman.

Whenever possible the ladder is pitched to the right of the window so that trapped people can be got out more easily.

Ladders being used for bridging — ladders, hose, and lines in use at the same time.

Photo: West Midlands Fire Service.

24

HOSE

Hose falls into two basic categories: delivery hose and suction hose.

DELIVERY HOSE, as its name suggests, is used to 'deliver' water to the fire. It has a non-porous inner lining of rubber or latex, and an outer jacket, nowadays normally made of nylon or terylene. Delivery hose comes in three thicknesses: 45mm diameter, 70mm diameter, and 90mm diameter.

SUCTION HOSE is used to *collect* water, either from a hydrant, or from open water such as a pond. 'Hard suction' hose is metal-reinforced and may be 76mm, 102mm, or 140mm in diameter, depending on the type of pump the fire appliance has. 'Soft Suction' hose is simply normal delivery hose that is being used to collect the water and feed it to the appliance. HOSE REEL HOSE is a form of delivery hose but is much thinner (only 19mm) and has no outer jacket as such. It is carried on a drum fixed to the fire appliance.

USING WATER FROM THE APPLIANCE ITSELF

For small fires, the water in the tank of the fire appliance may be enough. This just involves unwinding enough hose from the hose reel to reach the fire while another fireman switches on the pump and opens the valve to the hose reel.

But for larger fires, the water supply has to be increased, and a hydrant is brought into use. The following photographs show how this is done.

Delivery hose and hard suction hose in use on a water tender.

A fireman using hose reel equipment.

The pump is turned on.

25

1. A hydrant opened to show the valve.

*2. **The hydrant is opened with a special key, and a standpipe is fitted to the hydrant valve.***

3. Soft suction hose is run out ready for connection.

4. The hydrant is tested to make sure there's no blockage.

5. One end of the soft suction hose is connected to the standpipe.

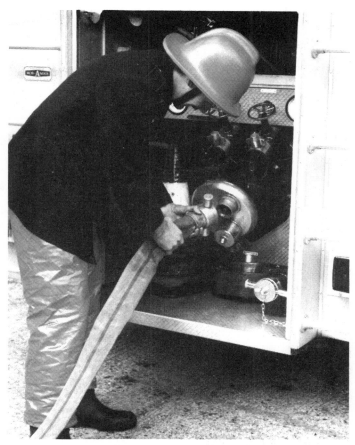

6. The other end is connected to the inlet of the appliance.

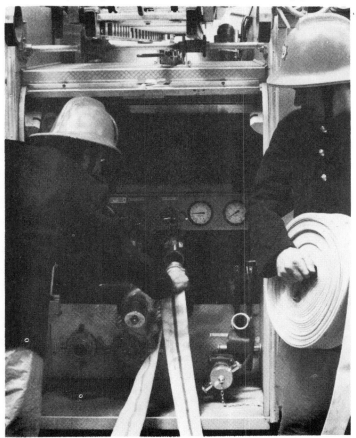

7. Next, delivery hose is connected to the outlet of the appliance. Another fireman runs out the hose.

8. The branch and nozzle are added.

9. *Both pump and hydrant are turned on. The fireman now has a steady supply of water.*

USING WATER FROM A POND OR STREAM

In cases where the water supply is some distance from the fire (e.g. in open country) firemen use any open water that happens to be available. The following photographs show how water is used from a stream.

USING WATER FROM A STREAM (8 stages)

1. *The equipment needed: 2 lengths of hard suction hose, strainer and basket, and line for securing.*

2. *The hard suction hose is coupled to the inlet of the appliance.*

3. Strainer and basket are secured to the far end of the hose to prevent leaves and other debris from fouling the pump. The equipment has been lashed to the appliance to reduce strain and unnecessary movement.

4. The hose complete with strainer and basket is lowered into the water.

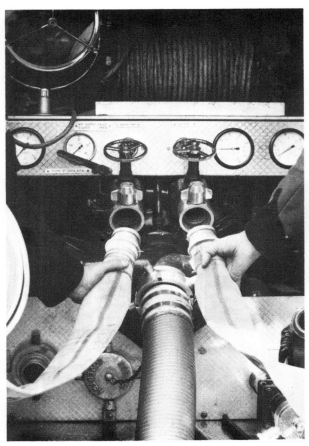

5. *Delivery hose is connected to the outlets of the appliance.*

6. *The hose is then run out.*

7. *Branch and nozzle are added.*

8. *Finally the appliance pump is switched on.*

As hose is charged with water for the first time, water forces its way at high pressure through the hose, causing it to 'snake'. The movement is powerful enough to knock a man off his feet. For this reason any fireman holding the branch as it's about to be charged must make sure he's in a safe position before giving the order "Water on" (see below).

Supporting hose as it is charged.

It is normally preferable to take hose to upper floors from *outside* the building. This is quicker and easier than dragging it through the building, which causes chafing and exposes the hose to damage from the fire itself.

First, a line is attached so that it can be hauled up (above right).

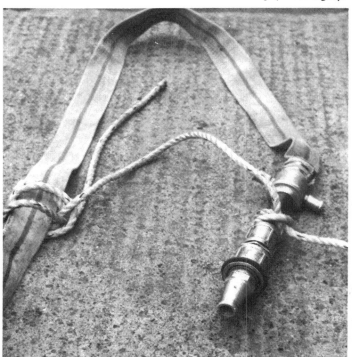

2. . . . . and then to the branch.

GETTING HOSE TO UPPER FLOORS USING A LINE. *(Sequence, 3 stages)*

1. First a line is attached to the hose . . . .

3. The hose is then drawn up to the required floor.

MAKING UP

When the Drill is over, the Officer in Charge orders the men to "Knock off and make up". The water is turned off, and everything is generally tidied up and replaced on the appliance.

Before the hose is 'made up', surplus water is drained out of it. The fireman lifts one end of it shoulder-high and passes the whole length through his hands. This is called UNDER-RUNNING. Later it will be hung to dry, either in the drill tower or in a special room with plenty of ventilation and heating facilities.

After under-running the hose is rolled up ready for stowing on the appliance.

There are several other ways of making up hose, but this is the most common.

Under-running a length of hose.

Hose being rolled before stowing on the appliance.

HOW IS FOAM MADE?

The process of making foam basically consists of introducing foam concentrate liquid into the water being pumped through the delivery hose to the fire. There are several points at which this can be done, by using certain special fittings. The following photographs show one of several methods of making foam.

MAKING FOAM. (Sequence, 3 stages).

1. The equipment needed: (from the back) Foam concentrate, foam branch, mixed foam generator (MFG 5A), pick-up tube.

2. Delivery hose is prepared in the usual way, but with the generator inserted between lengths, and pick-up tube coupled to it.

3. The rigid end of the pick-up tube has been put into the container of foam concentrate, thus introducing foam to the water before it leaves the delivery hose.

Here, foam has been used to extinguish fire in a crashed road tanker.   Photo: Essex County Fire Brigade.

A LOOK ROUND THE SMOKE CHAMBERS

As well as doing drills with the appliances, firemen have to do Breathing Apparatus Drills in special smoke rooms. First let's have a look at the rooms themselves.

Smoke rooms are built for a specific purpose: to provide an area where firemen can train in conditions similar to those they're likely to find themselves in at a fire. That's to say, in unfamiliar surroundings where the dangers are unknown, and in heat, smoke and total darkness.

To go into a smoke chamber is like walking into a tomb. No windows or fresh air here! The whole place smells of smoke and is lit only by a few light bulbs which are switched on while the area is prepared for the drill, and again later when it's over.

No two Breathing Apparatus rooms are the same; nevertheless they have certain features in common. A typical set-up includes a kind of obstacle course called the RAT RUN, which is really a maze of galleries constructed from open wood panelling similar to that used for cattle pens.

In places these galleries are of normal room height; in others, just high enough for firemen to slither through on their stomachs — as a rat would! Odd hazards are dotted about with planned cunning: metal rollers forming a cattle grid; a ladder with a few rungs missing; an unsafe floorboard or trapdoor — all waiting for the unwary! Sliding panels can be moved about to open or seal off various passageways, making it possible to arrange different layouts each time to keep the firemen guessing! Other obstacles include such niceties as a 6-metre length of sewer pipe to crawl through, or a vertical JACOB'S LADDER to climb.

By way of adding a little confusion, the officer in charge will make use of any odd object he can lay his hands on. A length of hose, for instance; a few metres of cable; items of equipment or household furniture. After scheming and setting up as many traps as he can, he'll send the crews in to negotiate the finished layout, and wait for them to appear at the exit point, knowing he's done a good job.

Let me remind you that they have to find their way through all this confusion in complete darkness, in heat and smoke, and with the added complication of having to wear Breathing Apparatus. Here in the smoke chambers firemen learn two important things — to work in cramped conditions and to expect the unexpected!

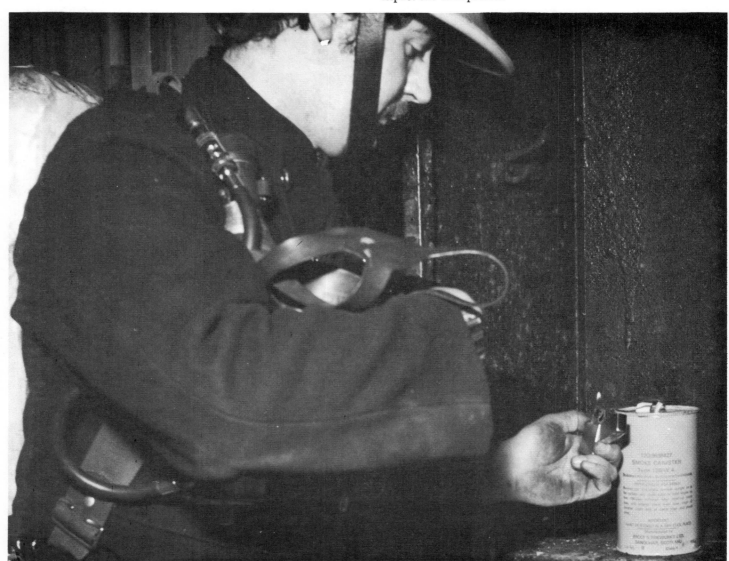

Lighting a canister of artificial smoke in readiness for a Breathing Apparatus Drill.   *Photo: Jaime Graham.*

At large fires a greater number of Breathing Apparatus teams is required, and the STAGE 1 Control Board is replaced by the larger STAGE 2. Here a fireman slides his tally into a slot in the Stage 2 board as the B A Control Officer (in this case a Sub Officer) prepares to add details.

Photo: Jaime Graham.

PREPARING SMOKE CHAMBERS FOR A DRILL

It takes 20 minutes or so to fill the chambers with hot air and smoke ready for the drill. There are several ways of doing this. In smoke rooms built of heat-resisting fire brick I've seen a collection of material — rags, timber, straw, old furniture, etc. — set alight and left to burn until the chambers looked like a kiln before the men were sent in.

Alternatively, a canister of chemical smoke can be used (such as is used in fireworks) while hot air is pumped into the room through vents (see facing page).

But the most popular method seems to be, by heating vegetable oil in a special machine called a SMOKE GENIE. The smoke is pumped out by pressurized carbon dioxide gas.

While this is being done, firemen taking part in the exercise test their Breathing Apparatus sets outside and each hands in to the B.A. Control Officer a tag called a TALLY. On this is written the fireman's name, the amount of air in his cylinder, and the time he's due out.

Each tally has its own key attached, to unlock each fireman's Distress Signal Unit if the need arises. The B.A. Control Officer attaches the tallies to his control board and jots down any relevant notes. It's his job to keep tabs on the rest of the team and make sure each man comes out by the time he's due. Finally, the officer in charge checks that the rooms are ready, turns off electrical systems and switches off the lights.

A BREATHING APPARATUS DRILL

When the officer in charge has given the men their instructions he sends them into the chambers in twos. For safety reasons firemen never enter a burning building alone.

The object of the exercise is most often to locate a 'body', or a piece of equipment, and bring it out. Sometimes the first crew in is given the job of laying out a guide line by which they can find their way back to the entrance. Also, firemen following later can run branch lines off it to search other parts of the building.

This guide line is an intriguing piece of equipment consisting of a 62-metre length of rope with pairs of tabs at 2½-metre intervals all the way along it (see below).

A Breathing Apparatus drill in progress in total darkness — but for the benefit of the cameraman without the usual heat or smoke! The slender guide line is a fireman's lifeline in a burning building. The position of the tabs (seen between the hands) tells the men they are going INTO the building.

Photo: Jaime Graham.

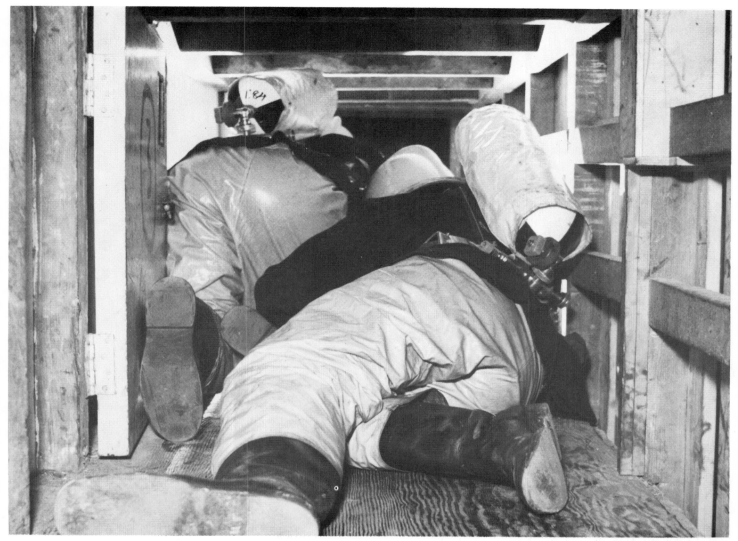

Firemen must learn to work in confined spaces. These men are training in the 'RAT RUN', and have entered a section of the crawling galleries via the open door whose hinges can be seen on the left. Another time it may be closed, altering their route.
Photo Jaime Graham.

The tabs themselves are 15cm apart, and as you can see from the photograph, one is knotted twice so that it is shorter than the other. As a fireman enters the building he fixes the line by the snap hook to a firm point outside, then pays out the line from a special container so that the shortest tab is nearest to him. On his way back of course it will be furthest from him, and this is how he'll know the way out, recognizing the tabs by touch.

When a fireman is inside a burning building among all the chaos of the fire and unable to see, this is his lifeline back to safety.

To the outsider a Drill such as this might seem easy, but it's one thing to take part in an exercise out in the open on the fire station yard, able to see and hear all that goes on, and quite another to be sealed off in darkness for half an hour or more, finding your way through the uncertainties of the smoke chambers.

Imagine for a moment what it's like for the fireman. He's already wearing his bulky fire gear when he dons his Breathing Apparatus, adding something like 14 kg to his weight. Now he straps the face mask over his head, and visibility through the heat resistant glass of the visor is considerably reduced. He also experiences an unpleasant 'shut-in' feeling.

As he makes his way into the pitch darkness of the smoke rooms, he hears the steady 'hiss-hiss' of air passing through the valves of his set. He shuffles forward, careful to test the flooring with each step, in case there's a sudden cavity or something that may cause him to stumble. He keeps his head down, anticipating low beams and other obstacles, searching all the time with the backs of his hands. He's been trained to do this so that if he touches electrical wiring exposed by fire, the automatic contraction of the muscles will snap his hand away from the current, lessening the risk of electrocution.

As he goes on, he loses all sense of time and direction, experiencing the feeling of being buried alive. However great the temptation to take off the mask, he must resist it. This is a feeling all firemen know and must learn to master, for to remove a mask in fire conditions is to risk losing your life. Somewhere ahead lies the exit point, and he must persevere towards it.

Crawling on hands and knees . . . . slithering along on his stomach pinned between floor and ceiling by the cylinder on his back . . . . climbing a broken ladder . . . . stumbling as he feels for a passageway to lead him to the outside . . . . At last he's there! Drained of energy by the heat, by his physical effort, and by the nervous tension of finding his way through an unpleasant environment, he can at last take off his mask and breathe all the fresh air he wants.

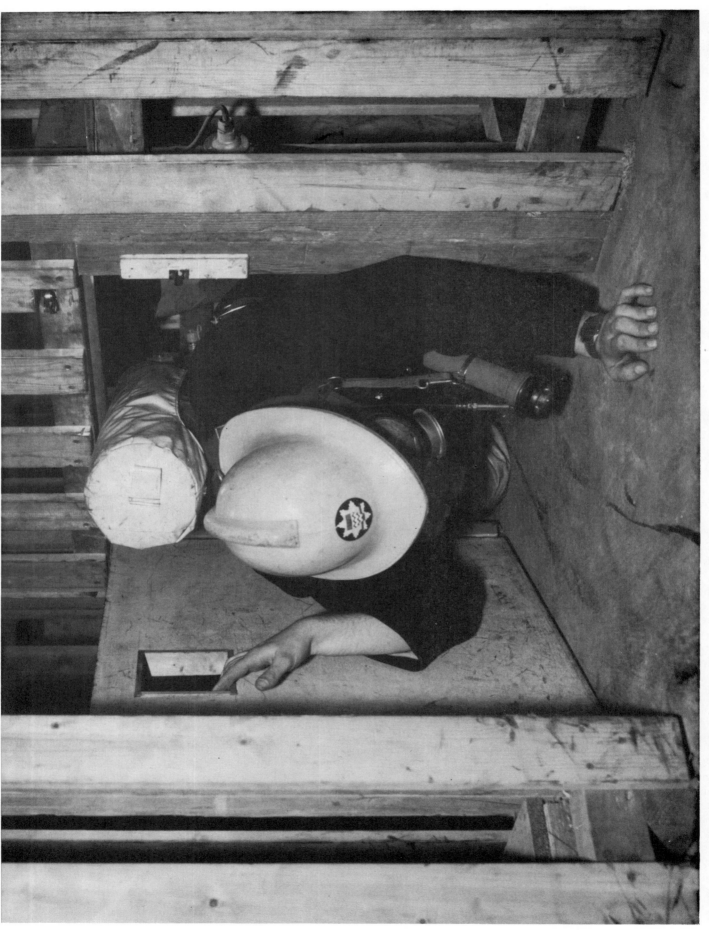

A fireman emerges from the crawling galleries. There is barely room to ease his Breathing Apparatus set through.

A crew on its way through a 4 ft diameter concrete sewer pipe, equipped with rescue line. Even in the absence of heat, the men's face masks are steaming up from exertion, blinding the wearers.
Photo: Jaime Graham.

The simulated rescue of a 'casualty' — in this case a 12-stone dead weight dummy.

Photo: Jaime Graham

Ascending a bolt-upright JACOB'S LADDER through a trapdoor in very cramped conditions.

Photo: Jaime Graham.

In the adjacent Control Room a Station Officer monitors the progress of his crews through the smoke chambers. Lights on the board above his head show the positions of the men.
Photo: Jaime Graham

SAFEGUARDS

Because conditions *are* unpleasant, certain precautions are taken to safeguard firemen while they're in the smoke chambers. If a fireman needs help he sounds his Distress Signal Unit. In response to this the officer in charge puts on the lights and switches on a large extractor fan that clears the smoke in a matter of minutes. And in a special monitoring room next door to the smoke chambers, a watch is kept on the men's movements. On a map detailing the area, small bulbs operated by pressure pads on the flooring as firemen walk over them will

light up, showing where a fireman can be found if he gets into difficulties. While he's training someone will keep an eye on him and he knows he's got nothing to worry about.

When a fireman turns out to a fire it's different. Something stored in an unexpected place may explode in his face. Fire may expose an unprotected lift shaft. A whole floor may give way suddenly and plunge him into an inferno on the floor below. A wall may buckle, bringing down tons of red-hot masonry.

Faced with the real thing, there's no such thing as safety.

During this period in the timetable the fireman trains continually with the equipment he is expected to use. He learns by practical experience which tool is the best to use in a given situation, and to this end, any vehicle banished from further service on the road will serve for practice purposes.

Since the fireman uses an extensive range of equipment and it would be impossible to include every item, here is a selection of typical rescue equipment.

1) AIR BAGS
These figure prominently in rescue work, being quickly and easily inflated from compressed air breathing aparatus cylinders to lift or move wreckage and other heavy obstructions, the deflated bag first being eased under (or between) whatever is to be moved. To protect the bag from jagged surfaces a salvage sheet is placed around it before it is inflated as shown in the photograph below. Made of nylon coated with protective neoprene, it is strengthened by inner webbing straps. While deflated it is only about 2.5 cm in height, enabling it to be used in very narrow spaces where other equipment would be too wide. Inflation begins when the control levers are moved forward, and can be halted as necessary by reversing them. Used in pairs as in the photograph, fully-inflated air bags can lift up to 2.032 metric tonnes to a height of 58 cm. Wooden blocks are wedged under the load at each stage of the lift to make it safe, not only here, but in all similar situations.

Air bags being used to lift a car clear of the ground. They are being inflated with compressed air from normal breathing apparatus cylinders, fed through a reducing valve operated by the fireman on the right. A salvage sheet protects the air bags from jagged surfaces, and wooden blocks are placed under the load at each stage of the lift to make it safe.

2) The CENGAR SAW (pronounced 'SENGAR')
This pneumatic hacksaw, which uses toughened blades, is especially useful for cutting through rigid structures such as steering columns, brakes, pedals, etc. Apart from the obvious advantage of speed, it can be used in confined spaces (e.g. the interior of a car) where normal movement of the arm is not possible. In a dramatic race against time this tool was used to cut through steel bars over second-floor windows during the fire at Woolworth's, Manchester, in 1979, enabling several lives to be saved.

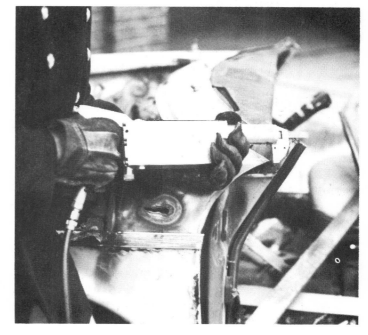

The Cengar Saw, a pneumatic hacksaw especially useful for cutting through rigid parts of a car and other solid structures. Protective goggles and gloves must be worn while using it.

A pneumatic chisel being used to cut away an area of a car's bodywork to expose the lock mechanism and thus gain access to people trapped inside when the door can't be opened in the normal way.
Photo: Hampshire Fire Brigade.

Oxy- propane flame cutting in progress on the hinges of a car door which has been made unopenable for training purposes. An asbestos blanket protects the occupant from the heat. *Photo: Hampshire Fire Brigade.*

3) The PNEUMATIC CHISEL

The Pneumatic Chisel is used to cut through thinner metal such as the bodywork of a car, as shown in the photograph.

It is quick and effective and requires little training. Note that here, as in the previous photograph, the fireman wears protective goggles and gloves.

4) OXY-PROPANE CUTTING EQUIPMENT

This equipment is reserved for cutting through heftier wreckage where other equipment is either unsuitable or inadequate to the task, since the naked flame poses the threat of ignition.

Oxygen and propane are fed from separate cylinders through a pipe to a nozzle with controls to adjust the mixture, which is then ignited. An intense flame is produced which rapidly heats the metal to melting point. To make a cut, the fireman operates a trigger to release a powerful jet of gas through the flame to blow away the molten metal. Metal up to 25 mm (1 in) thick can be cut by this method. Oxy-acetylene, although producing a faster cut, has disadvantages — acetylene is stored in larger, less portable cylinders and is less stable than propane, thus requiring greater care in handling.

5) HYDRAULIC RESCUE EQUIPMENT

Invaluable at road traffic accidents and other rescue situations, this equipment can be set up quickly, and since it works in any position, is particularly useful in confined spaces.

By repeatedly moving the pump handle up and down, hydraulic pressure is applied to the ram, which can be fitted with a variety of attachments to give a wide range of uses — e.g. lifting, pulling, pushing, spreading and clamping. A photograph showing this equipment in use appears in the chapter on road traffic accidents (see page 86). Here again, wooden blocks are used to shore up the wreckage.

Hydraulic rescue equipment in use. By moving the handle of the pump up and down repeatedly, hydraulic pressure is applied to the ram, which can be fitted with different attachments. Here, it is being used during training to spread railings so that the head of the 'casualty' can be released. *Photo: Hampshire Fire Brigade.*

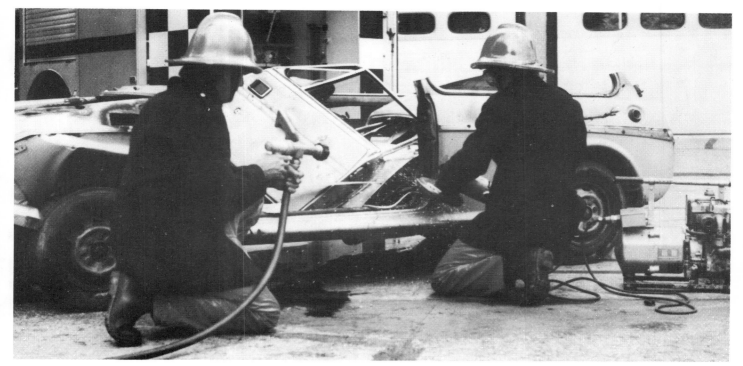

The fireman on the right is using a cutting disc powered by the portable generator at his side, to cut through the rigid metal of a car. A second fireman stands by with hose reel equipment ready to douse any fire caused by the sparks.

6) The CUTTING DISC

The fireman on the right is using a cutting disc powered by the portable generator at his side. This, like the Cengar Saw, is used for cutting through rigid metal, though not where people are in close proximity. The sparks given off by the cutting disc are a hazard, and a second fireman always stands by with a hose reel to douse any materials that may be ignited during its use.

Firemen position a ground monitor which will continue to direct water in a given direction, leaving them free to attend to other duties.

OTHER EQUIPMENT

The photograph above shows firemen positioning a GROUND MONITOR which will continue to direct a jet of water towards a given point, as on next page, leaving the firemen free to attend to other firefighting duties. Monitors such as this one can be used when a building is too dangerous for a fireman to approach, either because it has been weakened by the fire, or contains something that is likely to explode. Ground monitors are shown in photo on page 85 where they are being used to cool oil storage tanks. This is yet another purpose they serve, protecting buildings in the vicinity of the fire, by cooling them.

This photograph shows two types of ground monitors that have been positioned to saturate bulk quantities of burning paper.

Photo: Somerset Fire Brigade.

PORTABLE PUMPS

Portable pumps, seen in the photograph below, are an important asset to the fireman. Being light and easy to carry, they can be used in situations where it is not possible to use a fire appliance — e.g. on uneven ground or where access is limited. They are frequently used to pump out swimming pools or flooded basements. As well as being used 'solo' they are sometimes used in conjunction with the fire appliances, perhaps to draw water from a stream and pass it through to the pump of a water tender, or, again, as an intermediate pump between lengths of hose to boost the water supply on its way from the appliance.

Portable pumps being used to draw water from a stream lying directly behind the fireman.

Photo: Somerset Fire Brigade.

A RATHER SPECIAL PIECE OF EQUIPMENT

Before we move on to the Lecture period, let's take a look at one of the most versatile pieces of equipment a fireman uses. This unique item can be carried anywhere. It varies in weight and texture and can be formed into various shapes according to what it's to be used for — and its uses are many! Lifting and lowering, securing other objects, marking out particular areas, retracing routes, pulling down unstable buildings, hauling people to safety — these are all ways in which it can be used.

I refer, of course, to ROPE. In a fireman's hands it has a hundred uses. But before I find myself in hot water with the entire Fire Service, let me correct myself! A fireman, when speaking of lengths of rope, refers to them not as ropes, but LINES.

We've already seen lines in use during the hose and ladder drills, where they provided a means of securing other equipment, carrying hose to upper floors, and extending the ladders. Others appeared in the Breathing Apparatus drill — the hefty rescue line carried by the fireman in the sewer pipe, and the special guide line with tabs which firemen were using to find their way through the darkness of the smoke rooms. You will see other lines in use as we move on through the book.

Meanwhile you may be interested to see the variety of knots and hitches a fireman uses in the course of his work — perhaps even to experiment in making some of them yourself. You'll notice that some knots are made in such a way that the line passing through them can be adjusted in length. Others must be constructed so that they don't slip, whatever strain may be applied to them. This is essential to the safety of both the fireman himself and the person he rescues.

Knots and Hitches from the manual 'Fire Brigade Equipment'. By kind permission of the Home Office (Fire Service Inspectorate).

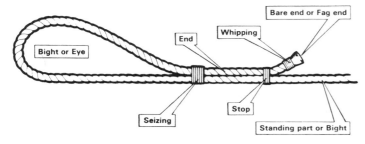

Terms used in describing bends and hitches.

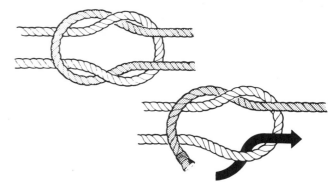

Method of making a reef knot. Used as a common tie for binding together two lines of roughly equal size.

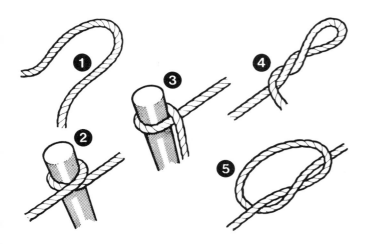

Elements of bends and hitches: (1) a bight; (2) a round turn; (3) a half hitch; (4) a twist; (5) an overhand knot. Most bends and hitches consist of a combination of two or more of the elements shown here. The overhand knot (5) is tied in a damaged length of hose to identify it, or to prevent a line from unreeving through a block.

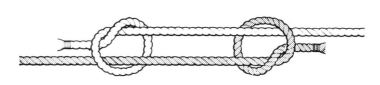

Method of making a fisherman's knot. To complete, each knot is hauled taut separately and then drawn together. Used as an alternative to the reef knot to join lines of equal thickness. Used by fireboat personnel.

Figure-of-eight knot. Used as an alternative to the overhand knot.

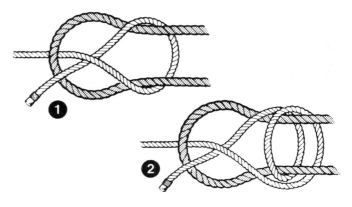

Method of making (1) a single sheet bend, and (2) a double sheet bend. Used to join lines of unequal thickness. The double sheet bend is used for extra security on wet lines or man-made lines.

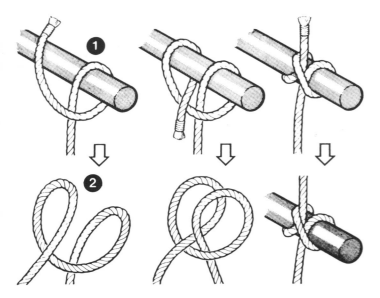

The clove hitch: (1) the clove hitch at the end of a line; (2) the clove hitch on the bight. Used to make a line fast to a spar, or to secure it to any object. Also to hoist fire brigade equipment as in next figure. Suitable for downward strain only.

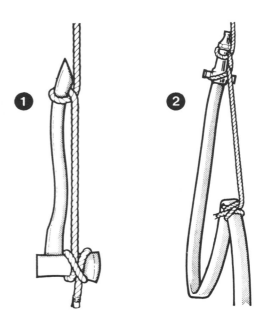

(1) Method of hoisting an axe using a clove hitch on the head and a half hitch on the haft. (2) Method of hoisting a length of hose and a branch using a rolling hitch on the hose and a clove hitch on the branch.

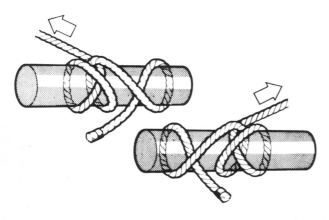

The rolling hitch, showing the direction of strain. A variant of the clove hitch, but can be constructed to withstand strain from left or right, as shown.

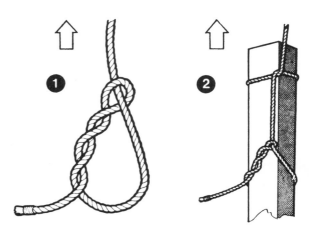

(1) Method of making a timber hitch. (2) Hoisting a spar by means of a timber hitch and a half hitch. Used to secure line for hoisting baulks of timber, planks or irregular objects.

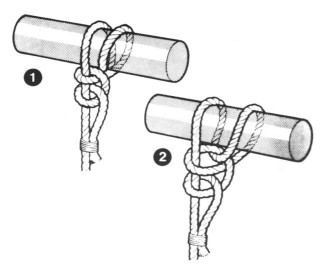

(1) First stage in the method of making a waterman's hitch. (2) The hitch completed. Used by fireboat crews to secure a skiff to the towing bitt of a fireboat.

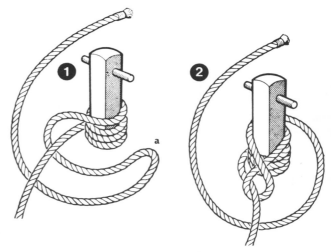

(1) A round turn and two half hitches. (2) The fisherman's bend. Both used to secure a line to a spar, ring, round object or another line. Fireboat crews use (2) to secure line to ring bolt.

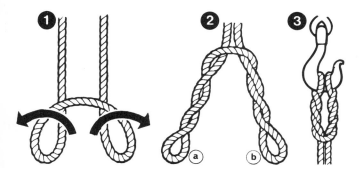

(1) and (2) Method of making a catspaw. (3) Line attached to a hook by means of a catspaw. Mainly used for attaching line to a hook.

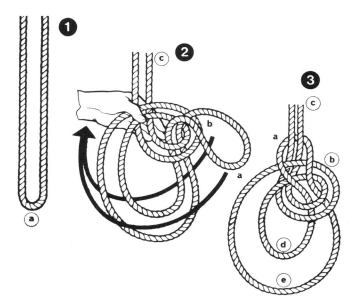

Bowline on the bight. Used as sling for rescue purposes. Bights at 'd' and 'e' are passed under knees and armpits respectively, of the rescued person.

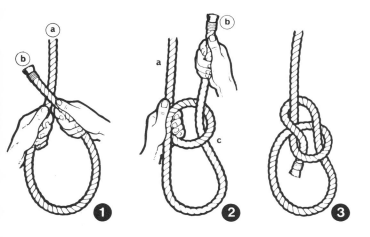

Method of making a bowline. Non-slipping noose for raising and lowering, or making temporary eye in a line. Also used to attach a line round waist when it is necessary to trail a line.

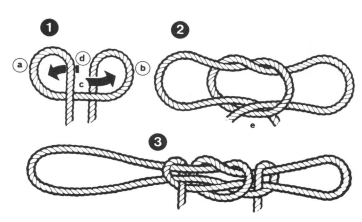

Method of making a chair knot. Alternative to bowline on the bight when a lowering line with legs is unavailable, or when makeshift stretcher is needed. One loop goes under knees, other under armpits.

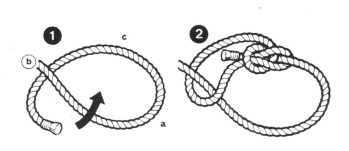

Method of making a running bowline. Can be put on ring, spar etc. Ends are secured so that noose can't be slipped over, by passing the line round the object, leading it under the 'standing part' and back, and tying bowline on the loop so formed.

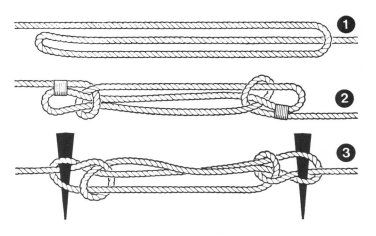

Method of making a sheepshank. Used to shorten a line without cutting. In figs. (2) and (3) the loops made by adding half-hitches have been made more secure from slipping by seizing with twine (2) and by wedging toggles in loop (3).

LECTURE SUBJECTS

The subjects a fireman studies at lectures are as numerous and varied as the incidents he goes to.

Fires of all kinds; accidents involving the release of trapped people; rescue work carried out in an infinite variety of situations; leakages of gas and dangerous chemicals — all these make demands on his skill and must be studied in detail. Let's consider just a few of these subjects.

EQUIPMENT

Equipment naturally figures prominently in lectures, and in many instances the lecture itself is closely linked with the Technical Training we have been discussing.

First and foremost of course, it's essential for a fireman to understand his Breathing Apparatus, as this is virtually a part of him while he's dealing with incidents. In dangerous environments his life depends upon it, so he must not only operate it correctly, but must be so familiar with it that he knows instantly if a fault develops.

He must have a thorough knowledge of everything he works with from day to day — his appliances, ladders and hose, pumps and extinguishers and — since water is an inseparable part of his work — water supplies and the operation of hydrants. He must also study special rescue equipment so that he knows what possibilities it offers at various incidents.

Equipment is dealt with in greater detail in the chapter about Technical Training, as you have seen.

COMMUNICATIONS

Communications — that is, the sending and receiving of messages — are absolutely indispensable in the Fire Service.

What's more since communications equipment has to meet the need for speed, it has become highly sophisticated. This in turn makes it costly, with the result that, next to wages, it's almost always the largest single item of expenditure in a fire brigade's budget. Because communications come into a fireman's work in so many respects, a working knowledge of them is essential.

For example he must be conversant with the various means by which members of the public can notify the brigade of emergencies, whether it be by the usual 999 call, by direct telephone lines or by automatic fire alarms.

He must also know how the brigade control room operates and be familiar with methods used to turn out fire appliances when a fire arises. The alarm may be given over a public address system, or by warning bells triggered by an emergency message typed by a teleprinter, or by radio alerters.

Most important of all, he needs to understand the workings of the VHF radio linking fire appliances, staff cars and the brigade control room, as well as the short-range radio sets used to pass messages at incidents, as he may have to use these at any time.

FIRST AID

A fireman's constant priority is saving life, so a knowledge of First Aid is essential. He is often involved at incidents where people need immediate medical treatment, either because they've been injured in an accident or are suffering from burns or the effects of smoke as a result of fire. In both cases there are likely to be unconscious victims, so he must know how to revive them, either by using mouth-to-mouth resuscitation, or by using a resuscitator. He may also have to give heart massage to bring the casualty back to life.

He must learn to splint broken bones and dress wounds of different kinds, from burns and scalds to freely-bleeding wounds from which a person would soon die if left unattended. He must be familiar with the method of placing casualties in stretchers and be capable of making a stretcher from his equipment when the real thing isn't available.

In addition to the normal medical boxes generally carried on the appliances, emergency medical boxes containing medical and surgical instruments are kept in reserve at certain fire stations and can be brought into use at incidents demanding on-the-spot emergency measures, provided that a registered doctor first authorizes it.

Firemen demonstrate method of resuscitation using automatic resuscitation apparatus. The 'casualty' is a member of the Eastern Electricity Board taking part in an exercise.
Photo: Suffolk Fire Service.

A workman with back injuries is brought to safety by firemen after an accident on the upper floor of a building under construction. He is strapped in a Neil Robertson stretcher — the same type that was used in rescue work at Moorgate.   Photo: Suffolk Fire Service.

Lowering by line by means of a stretcher (photographed during training). Equipment used: 13.5m ladder, section of short extension ladder, general purpose line, lowering line, and guy line. Photo: Hampshire Fire Brigade.

BUILDING CONSTRUCTION AND MATERIALS

To most of us, fire in one building is much the same as fire in another. The end result is destruction of property to a greater or lesser degree. We might therefore reasonably ask: "Why should a fireman interest himself in how places are built when more often than not all he sees of them is the ruins that are left?" But of course buildings are not all alike. They differ in shape and size, in the method of construction and materials used, and also (very important) in what they contain. It follows that they don't all behave in the same way in fire. Nor does fire behave in the same way in them. A fireman must therefore study building construction from a safety point of view. In an occupation as risky as his where virtually anything can happen, it is nevertheless in his interests to know what is likely to happen.

Look around you for a moment. What other buildings are there in your vicinity? Unless you live in a very remote area the number and variety will probably surprise you. Now remind yourself that a fireman must study them all!

He'll be interested in the tall office block with a lift shaft that can funnel fire from basement to roof; the old furniture factory up the road built largely of timber, with its dust-filled ducts along which fire can travel unseen; the church whose roof may drip molten lead on to him.

And he's almost certainly taken a good look at any especially large buildings in the neighbourhood such as warehouses and factories. In these there may be steel joists which, if fire reaches them, will expand and push out the walls. Or the roof may be supported by unprotected steelwork which, once it reaches a certain temperature, will collapse without warning, bringing down the entire roof.

And what about *your* home? Is it a simple terraced house or a country mansion? If it's a terraced house with the usual slate roof and you are ever unfortunate enough to have a severe fire upstairs, the fireman will have to watch out for the razor-sharp tiles that will shower down once the rafters have burned through.

If you live in a country mansion perhaps there are stone steps leading to the cellar? Solid, impenetrable steps, centuries old. But indestructible? — No!

Just suppose that one day a member of the family goes down to the cellar and is careless enough to drop a lighted cigarette end among some odd goods you store near the whisky barrels! With inflammable spirit around a fierce fire develops — so fierce that by the time you discover it the whole area is red hot. The stone steps too have heated up, and as the first rush of cold water from the hoses hits them they shatter completely — for this is a peculiarity of stonework. Apart from being in danger from falling masonry, if the steps were the only means of access to the cellar, the fireman is left with no way out. So you can see why Building Construction is such an important subject to the fireman. As outsiders we fail to appreciate that firefighting today is a true *science*.

A line being used to pull down an unsafe chimney following a severe house fire.   *Photo: Essex County Fire Brigade.*

An example of a roof whose rafters and battens have burned through, allowing tiles to fall.   *Photo: Essex County Fire Brigade.*

Fire can travel undetected through spaces under floors and roofs. These firemen are checking a ceiling void for fire spread.

Photo: Owen Rowland.

This photograph shows the effect of intense heat on brickwork. This has become dislodged, displacing the wooden beam.   Photo: Hampshire Fire Brigade

Note the overhead pipework here. Air Conditioning ducts and trunking carrying electric cables can also channel fire all over a building.   Photo: Owen Rowland

Here, walls have collapsed due to heat, leaving loose bricks lodged precariously at lower levels. Metalwork at the front of the building has buckled; so too has the steelwork of the roof, which can be seen hanging down inside the building.
Photo: James Clevett.

SYMBOLS AND MARKINGS

1) Hazard Warning Diamonds —

All chemicals are dangerous, especially when exposed to fire. Substances may then undergo chemical changes with the result that something even more dangerous is produced. Containers often burst in the heat, releasing different chemicals which intermix to form deadly liquids or explosive gases. This is also true of waste disposal sites, where dumped products may eventually come into contact with each other. This interaction between chemicals sometimes results in a spontaneous fire.

To keep him safety-minded a fireman is trained to look out for warning symbols, including those used in the labelling of products. He is likely to meet these in bulk on industrial premises, in storage areas such as shops and warehouses, and on vehicles used for the transportation of goods. (See also the Hazchem Code, explained in the chapter on chemical spillages.) Products bearing these warning symbols are found in our homes too, though in smaller quantities. Here are some Hazard Warning Diamonds commonly used on containers and container lorries (see facing page).

2) Biohazard Symbol —

Another warning symbol the fireman sometimes meets is the Biohazard sign used in hospitals and research laboratories where germs are being studied (see illustration in colour section).

3) Cylinder colour codings —

Cylinders, whatever they contain, present a lethal risk at fires. Unless they're kept cool, inside pressure continues to build up, finally causing them to explode. The force of a cylinder explosion is often equal to several hundred pounds of TNT, and results in extensive damage.

So that their contents can be identified, cylinders are painted to conform to the British Standards Institution colour code as shown in colour section.

Cylinders must now be stencilled according to their contents, but colour coding remains a valuable method of identifying them from a distance.

Canisters of flammable liquid bearing the hazard warning diamond, photographed at the roadside during an incident. Photo: Owen Rowland.

HAZARD WARNING DIAMONDS

FLAMMABLE LIQUIDS

FLAMMABLE SOLIDS

FLAMMABLE GASES

TOXIC GASES

NON-FLAMMABLE
COMPRESSED GASES

TOXIC SUBSTANCES

HARMFUL SUBSTANCES
KEEP AWAY FROM FOOD

CORROSIVE SUBSTANCES

ORGANIC PEROXIDES

OXIDIZING SUBSTANCES

SUBSTANCES WHICH IN
CONTACT WITH WATER
EMIT FLAMMABLE GASES

SPONTANEOUSLY COMBUSTIBLE
SUBSTANCES

EXPLOSIVE SUBSTANCES

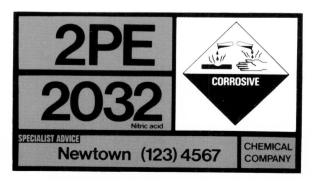

THE 'CORROSIVE' DIAMOND FORMING
PART OF A ROAD TANKER PLACARD

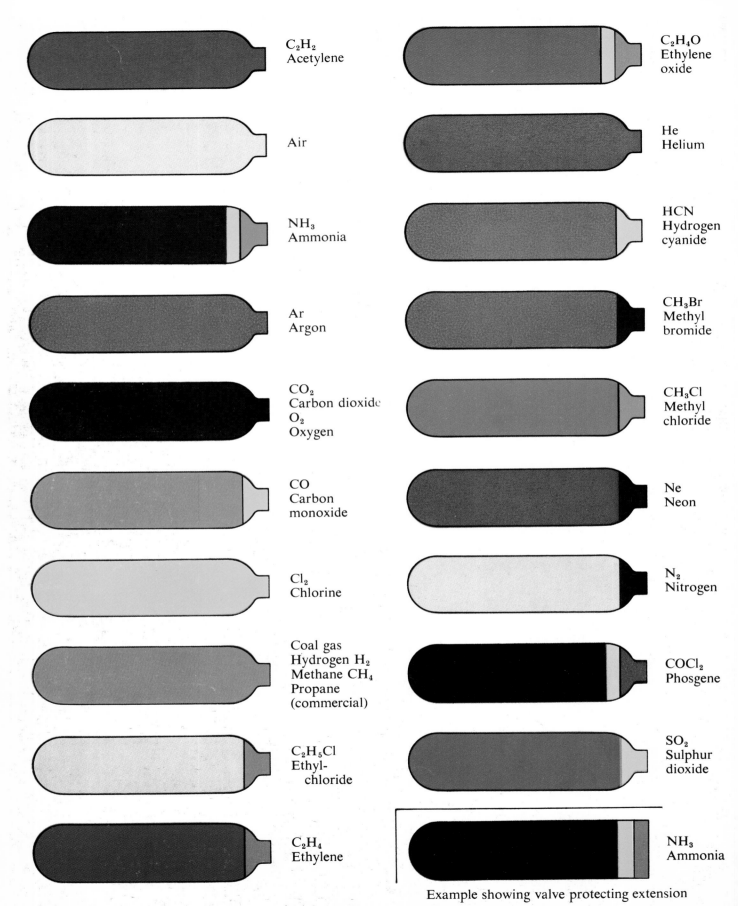

C_2H_2
Acetylene

Air

NH_3
Ammonia

Ar
Argon

CO_2
Carbon dioxide
O_2
Oxygen

CO
Carbon
monoxide

Cl_2
Chlorine

Coal gas
Hydrogen H_2
Methane CH_4
Propane
(commercial)

C_2H_5Cl
Ethyl-
chloride

C_2H_4
Ethylene

C_2H_4O
Ethylene
oxide

He
Helium

HCN
Hydrogen
cyanide

CH_3Br
Methyl
bromide

CH_3Cl
Methyl
chloride

Ne
Neon

N_2
Nitrogen

$COCl_2$
Phosgene

SO_2
Sulphur
dioxide

NH_3
Ammonia

Example showing valve protecting extension

A London office block ablaze from the third to the eighth floors.

Photo: London Fire Brigade.

A fire involving oil products at a waste disposal site, moving towards drums of dumped chemicals. This type of fire is often caused by the interaction of different chemicals. Photo: London Fire Brigade.

Kent firemen at work at a late-night head-on collision. A member of the crew supports the head of an elderly man dying from his injuries as colleagues work to free him. Photo: Chris Nelson.

The 'BIOHAZARD' sign a fireman may encounter in hospitals and research laboratories.

The emblem of the Fire Services National Benevolent Fund.

A photograph showing fire in straw. Like peat, it is difficult for the water to penetrate, and takes a considerable time to extinguish.

Photo: Chris Nelson.

An unusual Special Service incident — West Sussex firemen rescuing an injured woman from the cockpit of her glider after it had crashed into a tree and overhead power lines. A hydraulic platform was used to bring her to safety.
Photo: West Sussex Fire Brigade.

Burst containers among the debris of a warehouse used for storing chemical products, among them deadly Phenol which attacks the body's central nervous system when touched or inhaled. Several firemen were contaminated by this at the incident and needed hospital treatment. Only the brigade's speed in confining this fire averted a serious local disaster. Photo: Warwickshire Fire Brigade.

The photograph below shows some of the damage caused by an exploding propane cylinder involved in a small fire on a building site.

Arriving at the incident with the first fire appliance, the officer in charge was met by a workman, who told him that the fire was out. Luckily for everyone on the site, he went to check for himself, and spotting two cylinders* standing upright in a small fire, ordered everyone out of the building. Moments later as his crew were putting on protective 'anti-flash' clothing and were about the cool the cylinders with spray, one of them exploded, knocking the men off their feet and showering them with debris from the roof. All five were taken to hospital suffering from shock and various other injuries, but the fact that at the time they were carrying out their Station Officer's order to work from behind a pile of building blocks some 15 metres away, most probably saved their lives.

Windows of nearby houses were shattered by the blast, and fragments of debris were found embedded in a railway wagon a considerable distance away.

*One cylinder contained propane, the other, oxygen, but this was not known until later.

The remains of a cylinder that has exploded after overheating in fire. Photo: West Midlands Fire Service.

A brush with death .... The blast from an exploding Propane cylinder has blown out the roof and walls of this partly-built warehouse, damaging the appliance and injuring the crew, who were in the building at the time. They are seen here investigating their injuries a few minutes after the explosion.
Photo: London Fire Brigade.

TOPOGRAPHY

At frequent intervals firemen take an appliance out on a 'topography' run during which they tour the local area getting to know its general layout. As they travel around locating various roads and buildings, they make a note of any that present a particular risk or difficulty of access. Is there a danger from overhead power lines, for example? Where are the nearest hydrants? And is there open water near by that can be used if needed? Anything that can be learned about the district is useful when an incident arises. Speed of arrival is crucial to the outcome of any incident — in fact firemen are required to arrive at the scene of an incident within a specified time limit, according to what type of area they serve.

The table below shows the time limits for given areas:

Risk Category	Number of pumps for first attendance	Approximate time limits for attendance (in minutes)		
		1st	2nd	3rd
A Cities and	3	5	5	8
—	2	5	8	-
B Large Towns				
C — Small Towns	1	8 - 10	-	-
D — Rural Areas	1	20	-	-

Example: In area A, 3 appliances will attend the incident, the first two to arrive within 5 minutes and the third within 8 minutes, and so on.

There are various ways of storing topographical information so that it is readily to hand. For instance, all fire stations keep Route Cards detailing the quickest route from the fire station to the given address (see photograph below).

A more sophisticated aid being adopted by various fire brigades is the 'CAB READER', which, at the touch of a switch, gives topographical details and water supplies on a screen, enabling firemen to check them on the way to the incident.

The subjects we've touched on in this chapter are a mere handful from a very long list. Firemen must study a host of others ranging from Hydraulics to Fire Prevention; from Physics and Chemistry to Rescue Methods; from Electricity and its symbols to Fire Service Legislation; Equipment Used at Road Accidents; Explosives; Special Fires: In Aircraft; in Chemicals; in Radioactive Materials . . . .

The list is endless — and there's always something new!

This period in the timetable brings the morning to a close.

A fireman checks topographical details and water supplies from the 'CAB READER' on the way to an incident. The machine is easily operated and fits neatly into the rear cab of the fire appliance.
Photo: Tyne & Wear Metropolitan Fire Brigade.

The afternoon begins with the crew of one appliance setting out on fire safety inspections in the local area. One of their responsibilities is to visit factories and other buildings where there may be certain hazards. They take with them a Fire Safety Inspection Card to be filled in with information relating to the building (see below).

One side of this Fire Safety Inspection Card is given over to recording what is manufactured on the premises and any hazardous materials used in processes, the position of the nearest hydrants and the addresses of factory personnel who can be contacted for keys or other assistance.

On the other side of the card is a space to be filled in with a rough plan of the building and its immediate surroundings. The plan shows any details relevant to safety while firefighting, especially power systems and other danger areas. The quickest route from the fire station is also given.

A Fire Safety Inspection Card showing the type of information recorded by firemen during their visits to factories and other 'special Risk' buildings.

On Side 1. they list details of particular dangers in the building and provisions for firefighting.

1. 1. (d) INSPECTION CARD

	DATES OF 1. 1. (d) INSPECTIONS	P. D. A.	
	.....24th June 1978	Wr.T: 24	
CARD No. 24/15	....	Wr.L: 24	
	....	E.T.: 24	
MAP REFERENCE: 847916			

NAME AND ADDRESS OF PREMISES	Filla-foam Products Sedgeman Street Dormers End Bucks	TELEPHONE NO. Dormers End 12345
NAME OF CONTACT	Mr. J. Bond, 6 Thrush Ave., Dormers End, Bucks. Mr. P. Glover, Flint Cottage, Havermere, Bucks.	
OCCUPANCY	Expanded plastic foam manufacturers	
CONSTRUCTION	A range of one, two and three storey buildings, brick walls, asbestos cement roofs	

SIZE	240 Metres X 70 Metres		
		FLOW P. M.	PRESSURE
HYDRANTS	150mm O/S Warners, Sedgeman Street	1800	4 Bars
	150mm O/S Steelware, Sedgeman Street	1700	3.5 "
	150mm O/S Car Park, Farrow Place.	2500	4 "

OTHER WATER SUPPLIES	
HAZARDS	Large quantity of T.D.I stored in bulk tank room (81828 Litres)

NOTES   Wet and Dry Sprinkler installations
      Fire Service inlet for boosting supply
      Alternative supply from 1136500 litre elevated tank
      Electrical Sub Station Isolating switch for F.B. on outside wall.

WrT = Water Tender
WrL = Water Tender Ladder
ET = Emergency Tender   All appliances from Fire Station No. 24
PDA = Pre-determined Attendance; the fire appliances allocated to attend.
Hydrants: LPM = Litres per minute.
        Pressure given in BARS.
Hazards: T.D.I. = Toluene Di-Isocyanate — an extremely dangerous fluid used in the manufacture of polyurethane foam. It is flammable and poisonous.

Wet and Dry Sprinkler Installations; = Automatic fire extinguishing systems whose pipes are charged with water during the summer months, but emptied of water during the winter when they are in danger from frosts.

Electrical Substation: Isolating Switch = the switch cutting off all power so that firefighting can be done in safety.

A copy of the card is carried on the two 'First Attendance' appliances, and another on the Emergency Tender. A further copy is kept in the fire station Watch Room so that stand-by crews — both retained firemen and crews from other fire stations — have access to it. And finally, a copy goes to Divisional Headquarters.

This fire safety inspection card is therefore an instant source of useful information that enables firemen to deal with incidents on the premises concerned with maximum efficiency and safety. Outside duties also include fire prevention inspections in various buildings (schools and hospitals, for example). Here, firemen check that fire warning and extinguishing systems installed to protect the building and its occupants are working properly. They also make sure that no alterations have been made either to the structure itself or to what the building contains, or to fire precautions within it, that may jeopardize the safety of people on the premises.

On Side 2. they draw a rough plan of the ground floor showing any features that are relevant to safety.

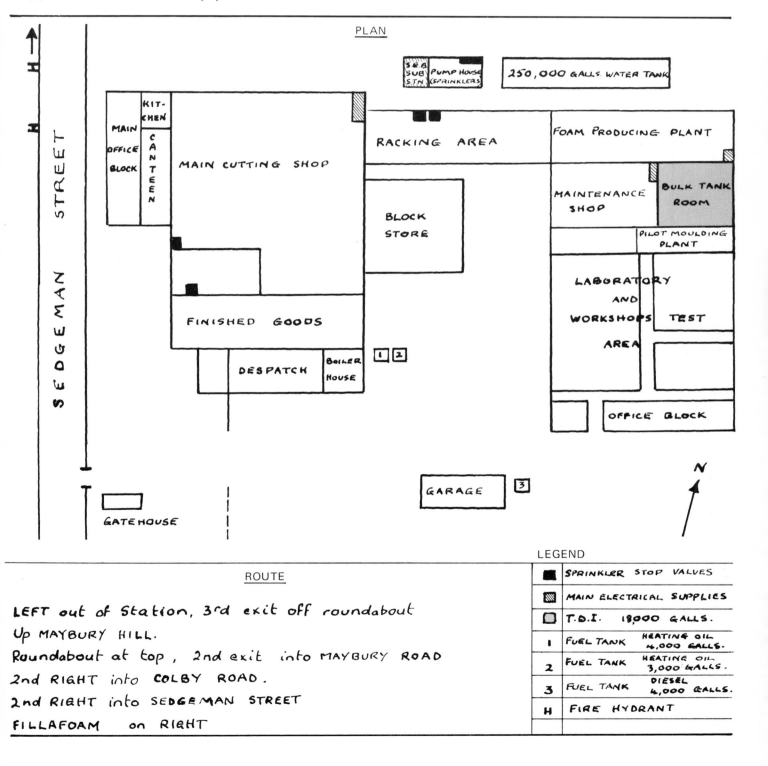

PLAN

S.E.B SUB STN

PUMP HOUSE (SPRINKLERS)

250,000 GALLS. WATER TANK

KITCHEN

MAIN OFFICE BLOCK

CANTEEN

MAIN CUTTING SHOP

RACKING AREA

FOAM PRODUCING PLANT

MAINTENANCE SHOP

BULK TANK ROOM

BLOCK STORE

PILOT MOULDING PLANT

LABORATORY AND WORKSHOPS TEST AREA

FINISHED GOODS

DESPATCH

BOILER HOUSE

1 2

OFFICE BLOCK

SEDGEMAN STREET

GATEHOUSE

GARAGE

3

N

ROUTE

LEFT out of Station, 3rd exit off roundabout

Up MAYBURY HILL.

Roundabout at top, 2nd exit into MAYBURY ROAD

2nd RIGHT into COLBY ROAD.

2nd RIGHT into SEDGEMAN STREET

FILLAFOAM on RIGHT

LEGEND

■	SPRINKLER STOP VALVES
▨	MAIN ELECTRICAL SUPPLIES
▢	T.D.I. 18,000 GALLS.
1	FUEL TANK HEATING OIL 4,000 GALLS.
2	FUEL TANK HEATING OIL 3,000 GALLS.
3	FUEL TANK DIESEL 4,000 GALLS.
H	FIRE HYDRANT

Off-station duties: Firemen testing water hydrants. This work is essential if the fireman is to be sure of a constant supply of water during a fire.

ON-STATION DUTIES

While their colleagues are out on fire safety inspections, the firemen remaining at the fire station will be getting on with various tasks such as cleaning the appliances, replacing equipment and re-stowing hose that has been dried, keeping paperwork up to date and so on.

In between all the other duties of the day there will be numerous visitors to accommodate — personnel from the Fire Prevention Department with queries or information needed by the Station Officer, staff from Headquarters with Fire Service business to discuss, deliveries of stores, etc.

As you can see, a fire station is a very busy place!

A party of visitors taking a look round the fire station.

On-station duties: Hose being washed to remove grit and other harmful substances such as chemicals which would cause it to deteriorate.

Report of a fictitious fire based on a genuine situation. This is one of the duties carried out by personnel on the fire station itself.

Report of Fire Barnshire

Brigade **99** Call No **399**

Tick broken boxes if applicable

Additional particulars to follow on FDR 2 · Yes · No ✓

Criterion number(s) if fire of special interest **85** **01**

1 General

1.1 Identification of incident: Date **31** **04** **82** Division **N** Stn ground **99**

Address 16 High Street

Barnville, Barnshire.

Name(s) of occupier(s) Mr. E. Jones.

Trade or business carried on. Hardware Shop

Risk category **C** Are premises certificated? No

1.2 Times:

Estimated time interval from ignition to discovery · Discovered at ignition · Short time i.e. under 5 mins · Fairly long time i.e. 5-30 mins ✓ · Very long time i.e. over 30 mins

Time of discovery **2359** Time of 1st call to FB **0005** Time of arrival of FB **0015** Time under control **0150** Time last appliance returned **0255**

1.3 Discovery and call:

Automatic fire detector

Fire discovered by: Person ✓ · Sprinkler · Heat · Smoke other (specify)

Method of call: Person via — 999 ✓ · Other tel · Running Call — Automatic via — Direct line · Central alarm · Exchange tel

Other (specify)

1.4 Further information Time of first call to F.B. delayed due to vandalised Public Telephone. Travel distance to second telephone ¼ mile.

2 Location of fire

2.1 Type of property where fire started: Shop

2.2 Fires in buildings only: Description of building

i) Detached · Semi-detached ✓ · Terraced · Other

ii) Single occupier · Multiple occ same use · Multiple occ different use · Under construction · Under demolition · Derelict · Unoccupied

2.3 Fires in buildings or ships: Details of location of fire

Floor of origin **G**

Place where fire started: On Roof · Other external structure · External fittings · Roof space · Room, cabin etc. ✓

(complete 2.6 next if necessary) (continue)

Use of room, cabin, roof space etc Hardware sales area

Openings: — Internal — None · All shut · Some open ✓ — External — None · All shut ✓ · Some open

2.4 Fires in vessels, caravans, trailers or outdoor plant only:

Dimensions in metres: Length Width Height

Materials of construction

2.5 Fires starting in road vehicles:

Make and Model Fuel Reg no

2.6 Further information Sales Area with many aerosol products on Sales Shelves.

3 Construction of building or ship

(if not applicable, tick box and complete Part 4 next)

`not applic.`

3.1 General

Is there any evidence of heat damage to the structure? ✓ `Yes` (continue)   `No` (Part 4 next)

Approximate year of construction `1960`

Number of basements `0`

Number of floors `2`

3.2 Approximate dimensions: (in metres)

	Length	Width	Height
Of room, cabin etc of origin	15	15	3
Of premises or ship	15	15	

3.3 Materials of construction and linings etc. directly affected by fire: (underline or specify ● as appropriate)

Roof/roof lining	Walls	Wall lining	Ceiling	Ceiling lining	Floor	Floor covering
asphalt/bitumen	asbestos		asbestos		brick/tile	
concrete	brick/stone	paint only	concrete	paint only	chipboard	carpet
felt	breeze block	paper	fibreboard	paper	concrete	linoleum
glass	chipboard	plaster	lath & plaster	plastics	earth	paint only
metal	concrete	plastics	plaster	timber	metal	plastics
plastics	fibreboard	timber	plasterboard		timber	
slate/tile	plasterboard		plastics			
timber	timber		timber			
●	●	●	●	●	●	●

3.4 Further information

4 Extinction of fire

4.1 Sprinklers and drenchers in area involved in fire: (if none, tick box and answer 4.2 next)   ✓ none

Installation: `Automatic` `Manual`   Number of heads actuated

Effect: `Did not operate` `Operated but did not control fire` `Controlled fire` `Extinguished fire`

give reason

4.2 Methods of fighting the fire:

Before arrival of Fire Brigade ... None

Methods used by Fire Brigade ... 2J from hyd via 1 WrT
2J from river via 2 WrT
12 C.A.B.A.

4.3 Attendance of Fire Brigade: (excluding relief attendance)

Name and rank of person in charge of 1st attendance ... L/Fm Smith.

Name and rank of person in charge of fire ... A.D.O. Pie.

Number of persons — Station Officer and above `3`

— below Station Officer `20`   Number of major pumping appliances `4`

Specify other appliances ... Emergency Tender, Control Unit

4.4 Further information ... Insufficient mains supply, pumping from river supply $\frac{1}{4}$ mile away

5 Damage and spread

5.1 Description of damage:

i) To item ignited first

P.V.C. Foam seating cushions destroyed by fire

ii) To room, cabin etc of origin

95% of Sales area of shop damaged by fire.

iii) Elsewhere on floor, deck etc of origin

Smoke damage to dry store in rear of shop.

iv) Elsewhere in building, ship, etc of origin

80% of living quarters on 1st. floor above Sales area damaged by fire, heat, and smoke.

v) Outdoor spread; spread beyond building, ship, plant, vehicle etc.

5.2 Total horizontal area damaged: (in square metres)

	Area damaged by direct burning	Total area damaged
In buildings or ships	185	225
Not in buildings or ships		

5.3 If any livestock were killed specify:

Species	Number	Species	Number	Species	Number
Dog	1				

5.4 Further information

6 Supposed cause of fire

6.1 Most likely cause: (if alternative causes are worthy of note, record these in 6.2)

Source of ignition   Electric Fire

Material or item ignited first (state composition of item)   P.V.C. Foam Seating

Defect, act or omission giving rise to ignition   Electric Fire inadvertantly left switched on.

Material or item mainly responsible for development of fire (state composition of item)   Aerosol and paint products.

6.2 Further information

7 Life risk

7.1 Involvement of persons: (as known to Fire Brigade)

Approximate number of persons at discovery of fire in:

i) Room or cabin etc of origin `0` ii) Other parts of building, vehicle etc `5`

Approximate total number who left the affected property because of the fire `1`

Approximate numbers of those who escaped by unusual routes

Fixed fire escape `   `   Ladder `   `   Through window `   `   Drainpipe, sheet rope, etc `   `   Climbing over roof, ledge etc `1`

Other unusual routes (specify numbers and routes) ....

7.2 Fatalities, other casualties and rescues: (complete one line for each person. Always refer to Guidance Notes for the codes to be entered in cols. 1-6)

Name	Age	1 Status	2 Location	3 Circs	4 Fatality/ Casualty	5 Rescued by	6 Rescue Method
A Mrs. E. Jones	35	F	C	D	F		
B Miss. E. Jones	16	F	C	D	F		
C Master E. Jones	12	M	C	D	F		
D Master D. Jones	8	M	C	D	F		
E							
F							
G							
H							

7.3 Further information ....

8 Explosions and dangerous substances

8.1 Explosions:

Explosion caused fire Fire caused explosion

Specify materials involved..... Butane Gas and Cellulose Paint
Specify containers involved..... Pressurised Containers and Aerosol cans.

8.2 Dangerous substances affecting fire fighting or development of fire:

Substance		Amount	Circs*	Effect on fire or firefighting	

*M = being made   S = in storage   T = in transit   U = being used

8.3 Further information.... Quantity of various paints, Strippers, and spirits based products and poisons involved in fire.

Signature .... XXXX *Smith* ....   Rank.... L/Fm ....   Date.. **1 May 1982**

FIRES IN VARIOUS ENVIRONMENTS

Fire happens literally anywhere, and in the following pages we'll be looking at various kinds of fires, among them some unusual ones, and the particular problems and dangers for firemen dealing with them.

Into the unknown .... A Breathing Apparatus team sets off into a burning factory.   Photo: Owen Rowland.

A chimney fire being extinguished from the roof with water from a hose reel.   Photo: Owen Rowland.

CHIMNEY FIRES

On the face of it, a chimney fire seems small and unimportant, but since it could lead to a more serious fire it's best not to neglect it.

Older houses, where the construction of the property leaves a lot to be desired, are the most vulnerable. When a chimney is first built, it is lined with a special cement called PARGETING. Separating this lining from the wooden rafters supporting the roof should be a protective layer of brickwork. As the property ages, the pargeting crumbles and falls away, and if the brick partition has decayed as well — or worse still, was never there in the first place — fire in the chimney may reach the rafters and set the whole roof alight. But if builders can be negligent, so too can the occupiers! For chimney fires arise when soot accumulates in the flue. In other words, when the occupier hasn't bothered to have the chimney swept regularly.

There are several methods of dealing with a chimney fire, but firemen prefer to deal with it from the grate upwards whenever they can.

First, they move everything away from the chimney and surrounding area — carpets, furniture, ornaments. Next, they spread a salvage sheet over the floor and make a 'dam' round the hearth from heavy sacking or some other suitable material.

Taking small amounts of water from a bucket, they pour it onto the fire in the grate. This creates steam which travels up the flue and helps to put out the flames. Pockets of fire within easy reach can usually be finished off with a stirrup pump, but if the fire is too far up, chimney rods are used to push the hose and nozzle up the chimney.

When a fire occurs near the top of the flue, firemen attack it by pouring water down through the chimney either using a hose reel hose or hand pump, or sometimes a bucket. Water must be kept off the hot chimney pots as the sudden cooling may cause them to explode. Apart from being hot, the pots may also be unsafe, so the fireman supports himself by holding onto the brickwork at the base of the chimney where it joins the roof. If the roof itself is unsafe, it may be necessary to bring in a hydraulic platform so that the fireman can work from the cage in safety.

It's sometimes difficult when working from the roof to find the right chimney in the first place, since one flue may lead down to several different fireplaces. Firemen and offending party alike would be unpopular to say the least if sooty water turned up unexpectedly in other families' rooms!

Some old houses have inspection doors in the chimney from which firemen can put out the fire if it can't be reached any other way.

Yet another alternative, as long as the chimney is sound, is to let the fire burn itself out, keeping firemen standing by in case they're needed.

A fireman never just assumes the fire is out. When everything seems safe, he makes a thorough check of the house in case it has spread. Finally comes the clearing up. All the mess is cleared away from the hearth. Carpets, furniture and ornaments are replaced and the room is left at least as clean as it was when he arrived. This is part of a fireman's professionalism — he takes pride in doing the job properly.

A fire involving the roof of a house.   *Photo: Huddersfield Examiner.*

FIRES IN HIGH BUILDINGS

How high is 'high'?

To the fireman it means any building with upper floors that can't be reached by normal fire appliances — usually those with more than six storeys.

Because of the rescue problems created by these buildings, they have to be carefully planned and constructed so that they're as fireproof as possible. As part of their safety precautions they must have built-in firefighting facilities and must be easily accessible from the road if a fire occurs. All such buildings have at least one RISING MAIN inside to provide a readily-available water supply. Depending on whether this pipe is kept permanently charged with water or is charged as necessary by the fire brigade, it is known as a WET RISER or a DRY RISER. Buildings over 61 metres high have wet risers, as fire appliances can't pump water at sufficiently high pressure at this height.

The main has outlets on each floor of the building, to which firemen can connect hose for firefighting, and it has several advantages over normal delivery hose. Not only can the riser carry a greater quantity of water, but it's also unlikely to burst. Another important advantage is that it minimizes the amount of hose needed, thus saving a lot of fetching and carrying as well as gaining precious time.

Nevertheless the fireman needs to get hose to the floor on fire as fast as possible, and to help him in this respect there is a *fire switch* close to the lift shaft which, once he has operated it, bypasses all other call signals at different levels, allowing him uninterrupted use of the lift while he brings in lengths of hose.

The firemen on the upper floors need to maintain contact with the officer in charge down in the street, so some form of communication link must be set up between them. Of the various systems available, (e.g. telephones, walkie-talkies, etc.) the best from the fireman's point of view are personal radios used in conjunction with the radio of an appliance at the incident, over which he can report on developments and ask for further help if necessary.

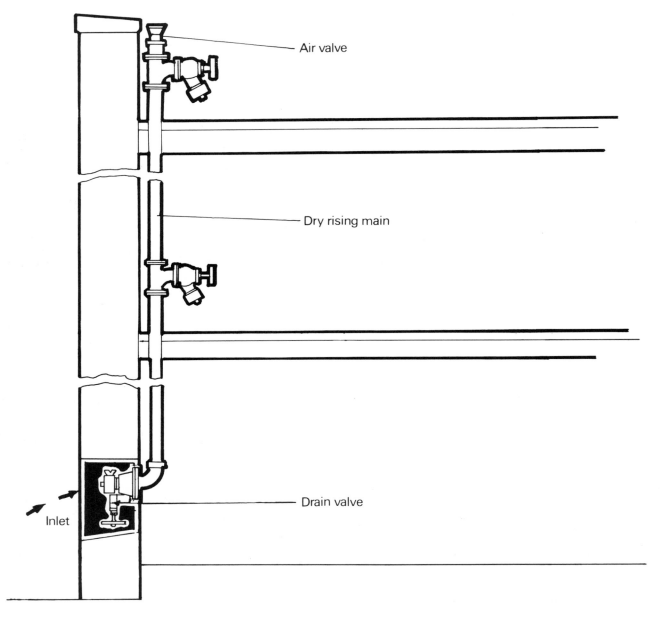

Diagram showing features of a dry rising fire main

By kind permission of the Home Office (Fire Service Inspectorate).

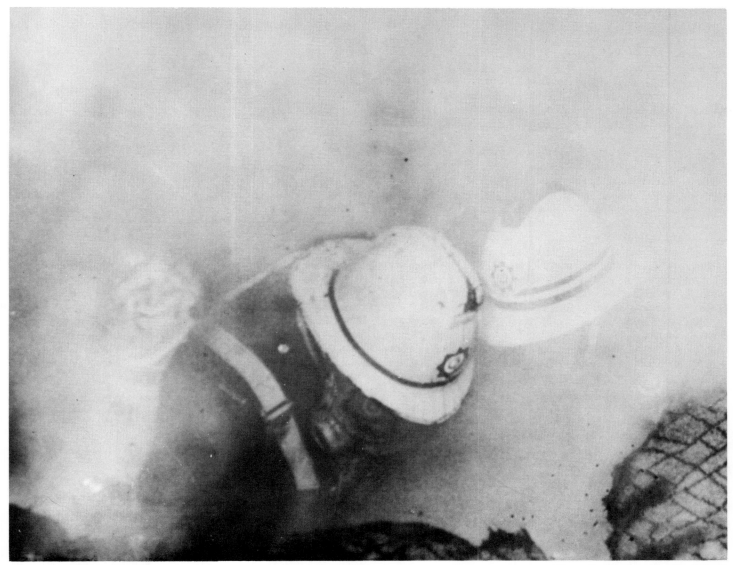

A photograph showing the build-up of smoke during a basement fire.

Photo: Buckinghamshire Fire Brigade.

BASEMENT FIRES

A basement fire presents the fireman with a particularly dangerous and unpleasant task.

First, access to the building is limited. This makes it impossible for him to calculate how far the fire has spread, and is a hindrance to operations generally. Finding his way about is difficult too, especially as he may have to pick his way through all kinds of stored goods. Worse still, to find the seat of the fire (its starting point) he must make a search below ground — a risky business when the building above may have been weakened by the fire and collapse on him, leaving no means of escape. A further problem — and an equally nasty one — is lack of ventilation. Heat and smoke, unable to escape in the normal way, build up until conditions become intolerable, and firemen are unable to work inside for long at a time.

Because of all these dangers the officer in charge of the incident will want to find out as much as he can about the basement and its contents before he sends his crews into it. Sometimes he is able to get hold of a plan of the building to study. Failing this, someone living on the premises may be able to give him a rough idea of the basement's layout.

Once a search has been made and the extent of the fire has been determined, men with jets are positioned to stop it from spreading. Then the basement can be ventilated to let out the hot air and smoke, by opening up cellar flaps and pavement lights. Firemen must make sure that when the flames push out through these openings, they don't enter open windows above and start fires on other floors.

Where there's a danger of fire funnelling up through the lift shaft to the upper floors, jets of water are trained on the base of the shaft to protect it.

If conditions are so bad that firemen can't get into the building, they'll have to extinguish the fire from outside, pouring in large quantities of water or high expansion foam. Foam is effective both in cooling the building and avoiding damage to its contents, but tends to break down in severe fire conditions, so the fireman may decide to use enough foam to clear a path, so to speak, then attack the heart of the fire with jets.

Flooding sometimes has to be done as a last resort, but this isn't always advisable, especially in large basements, as the considerable pressure of the water used to flood a large area might be strong enough to bring down unsafe walls. If flooding is done, however, as soon as the fire is out, the water must all be pumped away again.

FACTORY FIRES

The problems and dangers of fighting factory fires vary a good deal, depending on the construction of individual buildings and the goods produced there, so it's a little hard to generalize. A feature common to most factories, however, is the large undivided space needed for carrying out various processes, through which fire can spread very quickly, especially if the end products, or the materials used in their manufacture, are flammable.

A furniture factory, for instance, is a high fire risk for it contains wood shavings and dust, flammable polishes and supplies of timber, cloth, and interior fillings. A spark from an electrical power source, or friction from machinery, and there you have all the ingredients for an explosive fire.

Clothing factories are just as vulnerable, for here one finds dust (fluff), off-cuts of material, paper patterns and flammable stain-removers, all of which ignite easily. Here too, fire can spread rapidly, especially as many kinds of cloth burn fiercely. You may have noticed that both these industries produce fine residues, or dusts, which catch fire readily and spread fire across the surfaces on which they collect.

But did you know that certain dusts can cause an explosion? Sometimes dust particles get stirred up and, while they're drifting in the air, come into contact with a spark or some other source of ignition. A few particles will ignite first and explode, sending up a bigger cloud, which can then flash through a whole factory and devastate it, causing death and injury. Some industries are more prone to this danger than others, but perhaps the most obvious are those dealing in wood, paper and cloth, tobacco and detergent powders, and such foods as cereals, tea, and sugar, etc.

The recommended method of tackling fires on these premises is to first use a fine spray to avoid disturbing the dust until it has been damped down enough for jets to be used.

In factories the water used to put out the fire sometimes proves as destructive as the fire itself, partly because of the damage it can inflict on goods and machinery, and partly because its weight can cause weakened upper floors to collapse. But also, water is often *absorbed* by goods it comes into contact with, making them swell and become heavier. Once again the floors may become overloaded and give way. Where goods have been stacked against the walls of the building and become waterlogged, another danger arises – the expanding goods will exert pressure on the walls causing them to buckle and eventually fall. Firemen must keep a constant watch for bulging or cracks in the walls and keep well clear if they show signs of collapse.

Some of the goods that readily absorb water are fibres such as jute, hemp and sisal, grain, rags and newspaper. You can probably think of numerous others.

Stock can be protected from water damage by covering it with PVC-coated salvage sheets, and as long as the goods stand a

An intense fire involving rolls of jute and polypropylene carpet backing. The building collapsed within half an hour of the emergency call being received by Tayside Fire Brigade.

Photo: Dundee Courier and Advertiser.

Polyurethane foam on fire, giving off dense smoke. This fire burned so fiercely that all three buildings were gutted within ten minutes. Note the unprotected steelwork of the roofs which will collapse when it reaches a high enough temperature.
Photo: Dave Palmer

A five-storey mill manufacturing paper filters; showing the collapse of a section of the building owing to the levering action of falling floors and the uneven expansion of walls due to heat.
Photo: Greater Manchester Fire Service.

Where fire is concerned, nothing is sacred .... A 100-year-old church at Newport Pagnell burns to destruction. Note the steep angle of the roof, making it difficult to reach the fire without endangering the life of the fireman.
Photo: Ivor Leonard.

few inches clear of the floor (on pallets, for instance) sheeting will also help to eliminate absorption of the water and prevent overloading of the floors.

One other danger that the fireman meets in factories generally is high voltage electricity used to power machinery and tools. Some firms have their own substations where the current is converted down to suit their particular needs. During a serious fire, power cables and other installations may become exposed, and if a fireman were to brush against unprotected wiring while wearing wet clothing or while standing in water, or using a hose, he would be at great risk. Power must therefore always be cut off while he's working on the premises.

In addition to automatic sprinkler systems, many factories nowadays are equipped with automatic ventilating devices in the roof. These are panels which open up as soon as heat or smoke reach them, allowing the fire below to channel itself through the roof instead of spreading sideways through the building and involving a larger area. Apart from reducing the actual fire damage, this device also makes life easier for the fireman, since it prevents the building from becoming smoke-logged, allowing clearer visibility as well as better working conditions. Nevertheless, Breathing Apparatus is normally required while dealing with factory fires, because of the variety of toxic substances now in use in manufacturing processes.

A fire of suspicious origin breaks out in a derelict hotel in the early hours. More often than not such buildings are unsafe to start with, and working in or around them during a fire is very dangerous. This one was no exception — after a 12-hour battle involving 8 pumps, 2 Turntable Ladders and an Emergency Tender, the building had to be pulled down. Photo: Chris Nelson.

Woolworth's Store, Manchester, where the building's contents produced rapid firespread, resulting in the deaths of ten people. Goods on open display in any store are at risk if customers smoke on the premises, or if an electrical fault develops near a display.   *Photo: Greater Manchester Fire Service.*

Greater Manchester firemen in action — in Oldham a baby boy is snatched from his burning home.   *Photo: Oldham Chronicle.*

Tender arms lift a young child to safety, but it is already too late . . . . Six others died here at the Crypt Restaurant, Dover, including a fireman who was crushed to death by a girder when part of the building collapsed.
Photo: Chris Nelson.

HEATH AND GRASS FIRES

In rural areas, one of the fireman's busiest and most exhausting times is at the height of summer when the grass and scrubland of the open country have been dried crisp and heated by the sun to the point where a carelessly-dropped match or the hot exhaust of a car sets them exploding in flames.

With plenty of dry fuel in its path, the fire spreads fast and can soon cover such a large area that there are simply not enough firemen to contain it without outside help. Then, various volunteers join forces with the firemen, using spades and beaters to put out the flames. Where possible, natural firebreaks such as roads, streams, rough tracks and boundary walls are used to check the fire. Adjacent areas are soaked with water and any clear strips of land are widened by digging.

One of the problems of fires in open country is that the flames are spread by the wind and can change direction quite suddenly, threatening the firefighters. Fires of this kind create powerful eddies, often stronger than the prevailing wind, which again can drive the flames in unexpected directions.

A further problem is that uneven ground may make it impossible for fire appliances to get close to the fire. This, coupled with the fact that water supplies are often some distance away, means that water from ponds and streams may have to be used and be relayed through several pumping appliances, involving long lines of hose.

If the ground is firm and free of undergrowth, the fire travels over it without penetrating the soil below. If there's a covering of leaf mould or peat, however, the fire eats its way through it and travels underground, often as much as a foot or more below the surface. Although it's useful to soak the area with jets, the only way to get at this sort of fire is to dig out the soil and turn it over to extinguish the burning peat. Trenches have to be dug to stop the fire from spreading below ground, and if the fire is a large one, bulldozers may need to be brought in to do the job. The tasks of digging and beating, carried out amidst heat, smoke and flying ashes, leave the participants filthy, sweating and thoroughly exhausted.

Pockets of fire continue to burn underground without giving off much smoke, creating the misleading impression that the fire is out, so as a precaution the whole area has to be patrolled for hours afterwards to make sure there's no fresh outbreak.

A heath fire involving undergrowth and young trees.   *Photo: Hampshire Fire Brigade.*

This forest fire has so far involved low branches and undergrowth, but to the right of centre is just beginning to involve the crowns of the trees.
Photo: Hampshire Fire Brigade.

FOREST FIRES

Like grass fires, forest fires most often begin with *people*, and once begun, are not easily put out. Fuelled by undergrowth and in certain cases by sap and resin in the trees, the flames leap from tree to tree with incredible speed, overtaking anyone in their path. Once a sizeable fire has developed, the air currents above it become strong enough to counteract the prevailing wind and, as with grass fires, the flames can spread in any direction. Flying sparks are a danger too — they sometimes drift for up to half a mile ahead of the fire, and may start fresh fires behind the firemen, cutting off their retreat.

Large forests are split up into sections by rough tracks called RIDES, which firemen can use as firebreaks, clearing away undergrowth and trees alongside them and saturating the ground. If it's a large-scale fire, tractors or excavators may be brought in to plough up a strip of land. In a forest of fully grown trees where fire is passing through the crowns (tops) of the trees, the flames are out of reach, and a wide firebreak is the only hope of saving the timber. Beating is only effective on smaller trees and is carried out by firemen working in relays. As we've seen in the chapter on grass fires, smouldering fires in the leaf mould are halted by digging trenches.

At a forest fire a careful watch has to be kept so that firemen are not cut off by the flames. A rendezvous point is set up from which fire officers can keep in touch with developments by maintaining radio contact with others stationed at the limits of the fire, and give instructions as necessary. Throughout the incident fire officers consult with the Forestry Commission so that they can plan their attack on the fire.

Damping down continues for some time after the fire has died down and a patrol is maintained as a safety precaution.

PASSENGER AIRCRAFT FIRES

Almost all airports have their own fire brigades equipped with hefty appliances that can travel over rough ground and pump out king-sized quantities of foam to extinguish aircraft fires. When an emergency arises at the airport they're on hand straight away. Even so, local authority fire brigades are notified of all fires, and appliances are sent from normal fire stations on a pre-arranged system, to back up airport vehicles at fires and emergency stand-bys on the airfield.

Apart from this, quite often aircraft get into difficulties well outside the airport limits, and then the responsibility for firefighting and rescue falls to the local authority firemen. In all aircraft fires the overriding danger is the large amount of fuel on board which, if the fuel lines and tanks have been damaged, can flow freely around, ignite, and engulf the whole plane in a few seconds.

In a passenger aircraft, a considerable number of people may need rescuing, and this adds to the difficulties.

The fireman's chief concern is to protect the fuselage and its exits from fire and heat so that passengers can escape. This is achieved by spraying plenty of foam along the line of the fuselage, sweeping the fire outwards. The foam layer rapidly cools the surface and covers the flames. The metal primarily used in the aircraft construction is aluminium alloy, and if, as often happens, the fire has taken a severe hold before the firemen arrive, the aluminium may reach melting point (600°C) and drip to the ground. It then automatically becomes easier to cover with foam. Magnesium, on the other hand, which is also used in building of aeroplanes, burns so fiercely that the foam is destroyed, so water is used instead. Since magnesium and water react violently together, at first the burning intensifies, but as more and more water is added, the metal eventually cools down.

In such a fierce fire, the rescue of victims would be out of the question, but what if conditions were more favourable? Well, as you can imagine, forcible entry into a passenger aeroplane is no easy task, since the whole structure was designed to withstand different stresses during flight. Windows for instance, are super-strong and almost always too small to be used as a means of access. However, for rescue purposes certain window panels are made so that they can be removed — emergency hatches as we call them — and these are always clearly marked on the outside of the fuselage. There are also special 'break in' points where firemen can cut through the skin of the plane with hacksaws or power tools without fear of damaging fuel lines and other equipment. Again, markings painted on the fuselage show where these are.

For releasing passengers from seat belts, the fireman has a purpose-made 'quick release' knife, whose blade is specially shaped to avoid injuring the casualty.

Because danger is imminent, in some cases he may have to move a badly injured person against his own better judgement — he'll have to decide on priorities.

After being carried clear of the plane (uphill and upwind of it if possible) casualties must be laid on material of some kind to protect them from the ground, and covered to keep them warm, as at best they'll be badly shocked if not unconscious. As with all rescue work at this type of incident, the fireman works closely with the ambulance service, and will ask their advice as he goes along.

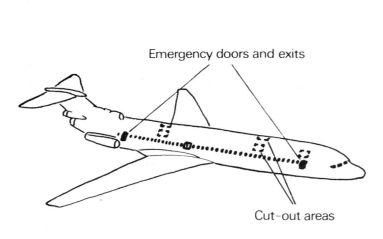

Method of emergency access to a typical passenger aircraft.

Typical 'break-in' markings on the fuselage to show where it is possible to cut through the airframe for an emergency entrance.

(Reproduced from the Manual of Firemanship Part 6b by kind permission of the Home Office.)

What a fireman is expected to know about aircraft accidents would fill a book! Apart from general procedures such as we've talked about, he must know about the parts, construction and layout of various kinds of plane (including helicopters and military aircraft) fuel systems, hydraulic systems, hazards associated with particular types of aircraft, seating arrangements, ways of gaining access, stand-by procedures and a host of other things besides.

For this reason, firemen are given every opportunity of making visits to airports and keeping their information up to date.

The streets of London .... A fire among rubbish.   *Photo: Owen Rowland.*

A six-pump fire involving four coaches of a train at Victoria Station.   *Photo: Owen Rowland.*

When fire broke out in this overnight sleeper just outside Taunton, twelve people died, trapped in its blazing corridors. Luckier passengers smashed windows and climbed out.
Photo: Somerset Fire Brigade.

A spark from a tanker's exhaust is thought to have triggered off this spectacular fire in a Lincolnshire oil well.
Photo: Lincolnshire Fire Brigade.

A deep-seated fire in the hold of this cargo ship kept Avon firemen busy for 20 hours. Ship fires pose special problems — a ship's layout is an unknown quantity, often making it difficult to trace the source of the fire. Working room is limited and the fire may be inaccessible. Also, the water used to extinguish the fire affects the stability of the vessel. Photo: Avon Fire Brigade.

A view giving some idea of the amount of hose required at the incident. Photo: Avon Fire Brigade.

A house severely damaged by fire following a gas explosion. Note the windows blown out by the blast, and slates dislodged from their battens.
Photo: Essex Fire Brigade.

After a gas explosion in this flat in which a woman died, firemen evacuated other residents while they dealt with the fire. The gas cooker had been turned on all night unlit, and the occupant, not noticing the smell of gas, died as she was lighting it next morning.
Photo: Greater Manchester Fire Service.

Here, an oil storage tank undergoing cleaning has become involved in fire and exploded, killing three workmen, two of whom have been buried by the collapsing blast wall. Owing to the unsafe condition of the wall the fire was allowed to burn itself out under supervision, with monitors positioned to cool the tank. Debris was removed by digger and crane in order to recover the bodies of the two men, the third being recovered by a two-man breathing apparatus team from within the tank when conditions were safe enough.
Photo: Grimsby Evening Telegraph.

ROAD TRAFFIC ACCIDENTS

Emergencies other than fires come in the 'Special Services' category, and this can include literally anything.

Top of the list for keeping the fireman busy, however, are road traffic accidents which, as well as demanding skill and ingenuity with his equipment, also involve him in a less pleasant aspect of his work. This is where he saves most lives, for though he goes to more fires than road accidents, far fewer people die in fires than die on our roads.

His overriding task here is to free people trapped in wreckage, many of them suffering from severe, sometimes fatal, injuries. Whatever a fireman's natural human feelings are when he sees a distressing incident like this, he can't give way to them. If he's to be of real assistance to people who desperately need it, he must get on with the job and work quickly and purposefully as always, taking care not to cause further injury to the casualty.

In the absence of outside help, he'll use the First Aid he's learned to keep a person alive and as comfortable as possible as possible until assistance arrives. This may involve giving the kiss of life, heart massage, stemming any bleeding and so on. But once the ambulance is on hand he passes the casualty on to the experts. In fact a fireman avoids all but the most necessary handling of victims, for by moving them he may do more harm than good, and in cases where a person's body has been pierced by some sharp object during the impact of the crash, it's usual for him to cut the person free of the wreckage without removing the object. Take, for example, the driver whose chest has been punctured by the steering column of his car. To remove it would leave a large wound from which he may bleed to death. Better by far to cut the column free of the car without further disturbance to the casualty and send him to the hospital where there are fully trained staff and appropriate facilities to cope with the emergency.

It's not unknown for firemen to accompany a casualty to the operating theatre, since they are equipped with various special tools and may be asked to assist with them before the surgeon can operate.

No two road accidents are identical, and circumstances alter the way in which the fireman deals with them. The following photographs provide some typical examples of accidents that are happening virtually every day on our motorways and country roads.

Hydraulic equipment in use at a road traffic accident — in this case to force apart the doorway of a lorry driver's cab in order to free him.

Photo: Owen Rowland.

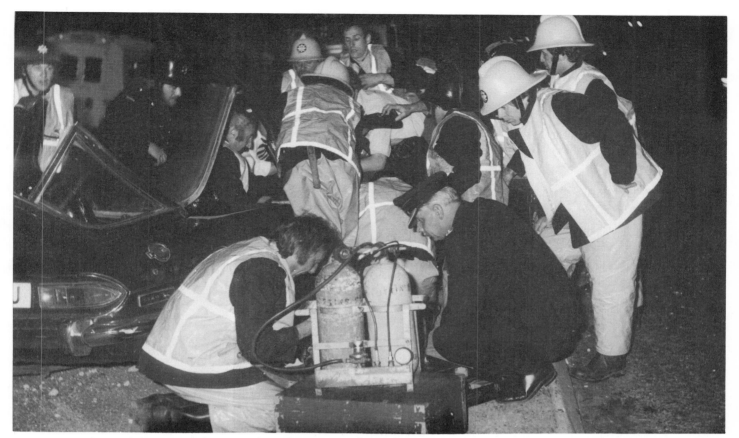

Oxy-propane cutting gear being prepared for use to release the occupants of a wrecked car.   *Photo: Owen Rowland.*

Firemen about to begin the task of releasing passengers from the upper deck of a tourist coach following a fatal collision with a low bridge.

Photo: Avon Fire Brigade.

Powerful 'alligator jaws' in use on the wreckage of an overturned car. Hydraulic pressure is used to force the jaws apart and prise open the wreckage.
Photo: Somerset Fire Brigade.

Wingham, Kent, where a car has been struck sideways on by an articulated lorry while pulling out at a junction. Firemen are about to release the dead driver.
Photo: Chris Nelson.

A lorry loaded with rolls of paper has gone out of control and ploughed into the front of a public house, leaving the building propped up only by the cab of the lorry. Firemen commandeered a crane held up in the resulting traffic jam to assist with rescue operations, and after removing much of the debris by hand, used a Tirfor winch, hydraulic gear and Cengar saw to release the driver, whose only injuries were bruised knees and legs. The rescue operation took 2 hours.

Photo: Lancashire Evening Telegraph.

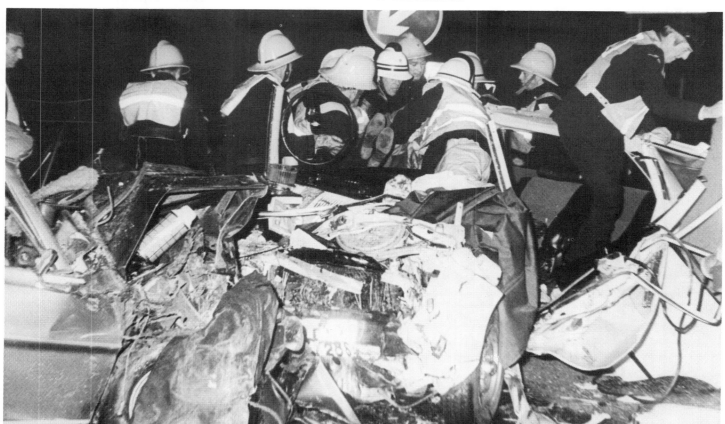

Moment of release for one of six casualties trapped in the wreckage of a head-on collision. Three died in this accident, which occurred at 11.30 p.m. on the A2 London to Dover trunk road when the driver of the car on the left missed the 'KEEP LEFT' sign and travelled down the wrong side of the road.

Photo: Chris Nelson.

Where to begin? East Sussex firemen begin the seemingly impossible task of locating and freeing passengers trapped in the wreckage of a late-night train crash. Three people died in the accident when one train, waiting at a signal during an emergency power shut-down, was struck from behind by another. Work at the incident, involving sixteen hours' non-stop use of oxy-acetylene cutting equipment, was finally completed two days later.

Photo: East Sussex Fire Brigade.

Fire crews arrive to free the bodies of four passengers from the remains of a light aircraft that crashed soon after take-off from Lydd airport.

Photo: Chris Nelson.

Rescue work in progress alongside the River Tay, where an express has rammed the back of a broken-down train. Several coaches plunged down the embankment into the deep mud of the river bed, making rescue work treacherous. Firemen helped rescue over 50 people at the incident, and are seen here using ladders to bring some of the casualties to safety.   *Photo: Dundee Courier and Advertiser.*

A Neil Robertson stretcher being used to winch a man to safety after being rescued from a sewage main under construction. Note: This type of stretcher was used extensively at Moorgate, being particularly useful in confined spaces.   *Photo: Hampshire Fire Brigade.*

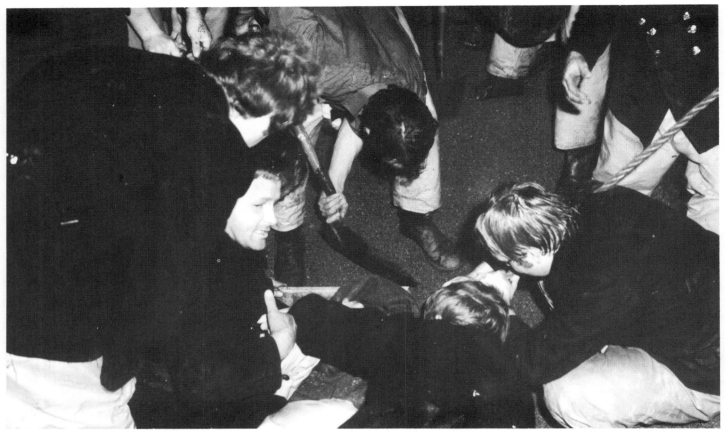

When it's a matter of life or death . . . .   Photo: London Fire Brigade.

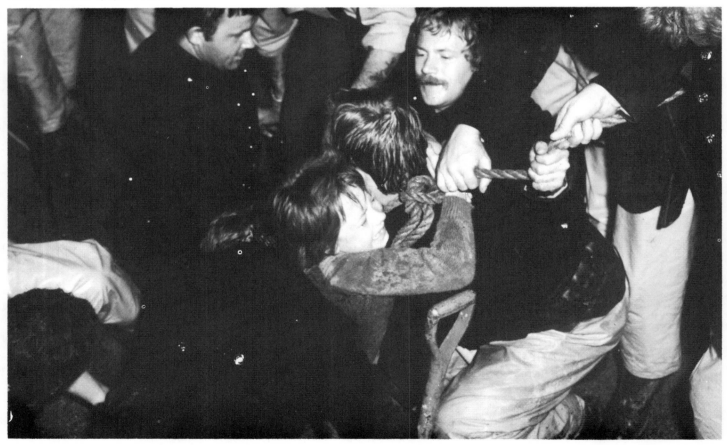

. . . . firemen are the best friends you can have. Playing on a lorry-load of gravel almost cost this boy his life. His friend pulled a lever — and the load was released through the lorry's funnel-shaped base, sucking him into it up to his neck.   Photo: London Fire Brigade.

The vertical access shaft to underground pressurized workings where a flash flood has blocked the doors, trapping 11 workmen tunnelling for a new sewer complex.
Photo: Greater Manchester Fire Service.

The scene underground, where firemen are about to pump out flood water before assisting the men up to the surface by ladders.
Photo: Greater Manchester Fire Service.

A mobile crane working at Tilbury Docks has overturned, plunging jib-downwards into 40 ft of water, crushing the cab against the quayside with the driver trapped upside-down inside it. The crane, which was balanced precariously, had to be made secure before rescue could be attempted. The trapped man, not seriously hurt, was eventually released after a combined rescue operation involving Essex Fire Brigade, a tug, a large mobile crane, and a floating crane provided by the Port of London Authority.   Photo: Essex Fire Brigade.

The driver of this low-loader was killed when the jib of the digger he was transporting (seen in the background) struck an overhead footbridge, bringing down one of its concrete sections on to his cab. Bucks firemen took an hour to release his body from the vehicle.   Photo: Milton Keynes Gazette.

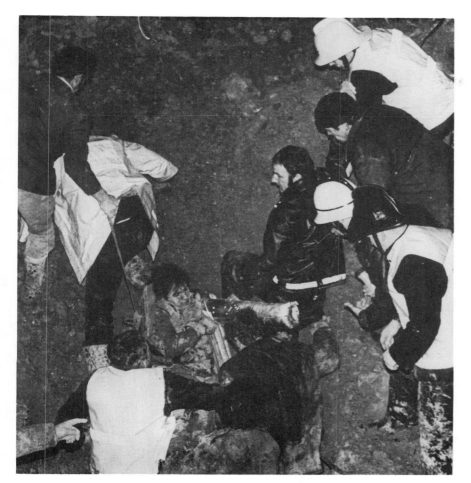

Buckinghamshire firemen spent over six hours freeing two men and the body of a third, buried in this 30ft deep trench by a landslide of excavated clay. One was rescued within the hour but the second, seen here being lifted to safety by two 'human chains', was trapped by the ankle under steel piling for 3½ hours. Firemen used a propane torch and hacksaw to sever the piling, cooling the immediate area with spray. The body of the third man was recovered 3 hours later.
Photo: Milton Keynes Gazette.

These firemen are faced with the unusual problem of rescuing seven cows that have played 'follow-my-leader' into a slurry pit. A JCB digger, at times in danger of becoming submerged, was eventually brought in to dig out a channel to each animal.
Photo: Somerset Fire Brigade.

THE MOORGATE DISASTER

It was the morning rush-hour on the last day of February 1975. All over London, buses and trains were packed with commuters making a last-minute dash to get to work by 9 o'clock.

Seventy feet underground a train was approaching the terminus at Moorgate, and passengers were ready to get off. But the train had already passed the crowds of people on the platform and showed no sign of stopping. It hurtled on at full speed into the short tunnel and demolishing the buffers there, smashed into the concrete wall beyond.

At a few minutes to nine, as firemen were coming off duty, the operations room at London Fire Brigade headquarters received news of the disaster from Scotland Yard. The message was promptly passed on to Stratford Control Room whose responsibility it was to mobilize fire appliances in the area north-east of the Thames, and the predetermined attendance of 3 pumping appliances and an emergency tender were accordingly sent to the scene.

The impact of the crash had put out the platform lights as well as the lights on the train, and emergency lighting would have to be set up before rescue work could begin. It was therefore only after closer investigation that the officer in charge realized that almost half the train had been wrecked in the tunnel, and that he had a major disaster on his hands. Without further delay he put the Major Accident Procedure into force, calling on all the resources of the emergency services.

Pending the arrival of ambulances and other fire appliances with special equipment, the first firemen on the scene began giving first aid to dazed casualties wandering about on the platform, and went to calm victims who were trapped in the wreckage at the front of the train. Some of them were barely visible among the debris and mutilated bodies, and in the chaotic, suffocating darkness, it was hard to tell what had actually happened.

In fact, the first coach had been forced up the concrete wall on impact and had folded in the middle, killing many passengers. As the second coach rammed it from behind, the first coach was simultaneously shunted back through the second, resulting in shocking loss of life. This also caused the roof of the second coach to peel up over the first, blocking the tunnel up to the ceiling. The third coach in turn had ridden up over the back of the second and was perched over the wreckage, jammed at an angle against the tunnel roof. The only space in which to work was the narrow strip between the train and tunnel side, and this was partly blocked by debris, presenting rescuers with a daunting task.

By soon after 9 o'clock all three emergency services were on the scene and had set up control units side by side opposite the nearest station entrance. Medical teams from St Bartholomew's and London Hospital set up first aid posts, one in the platform foyer and one at the end of platform 10. At the second of these there was provision for dealing with more serious injuries before the casualties were taken to hospital, and this no doubt saved many lives.

Within the first hour, with all three emergency services helping with the rescues, 25 people had been freed. Firemen were working like beavers, using every available tool to cut through the wreckage and reach the injured. It was really a job for oxy-acetylene work, but for safety reasons this couldn't be used in the presence of trapped survivors. Even now the heat from the floodlights and other forms of lighting being used was building up in the confined space and was soon over 100°F. Firemen stripped to the waist and were given salt tablets by the doctors to counteract the heavy loss of salt caused by sweating. Eventually because of lack of room and stifling atmosphere, the rescuers were cut down to a mere handful — a few firemen with cutting gear, and several medical staff with essential drugs and life-saving equipment.

With the third coach at last empty of casualties, work began on freeing passengers from the remains of the two front coaches. To make access possible, firemen cut an opening from the third coach into the second at the point of contact (the third coach was propped up over the end of the second) and another in the roof of the first coach, behind the driver's cabin.

Firemen found there was barely room to pass each other, and eventually a 'one way' system was adopted. This meant that they would enter the doorway of the third coach from the platform, make their way through the emergency opening into the second coach, drop into the tunnel through the far door, then climb along the tunnel plates and enter the first coach via the hole in the roof. Casualties were then strapped in special body-fitting stretchers and passed from one fireman to another in the nearside tunnel space, being transferred to normal stretchers on reaching the platform.

The police meanwhile had been fully occupied co-ordinating all the emergency services and clearing vital areas. They had also set up an information bureau about the passengers at

Relative position in which coaches came to rest (Side elevation)

The jacknife section of roof

Holes cut

Double doors

Movement of personnel, equipment and casualties (Plan view)

Holes cut   Tunnel mouth   Platform

Out   In

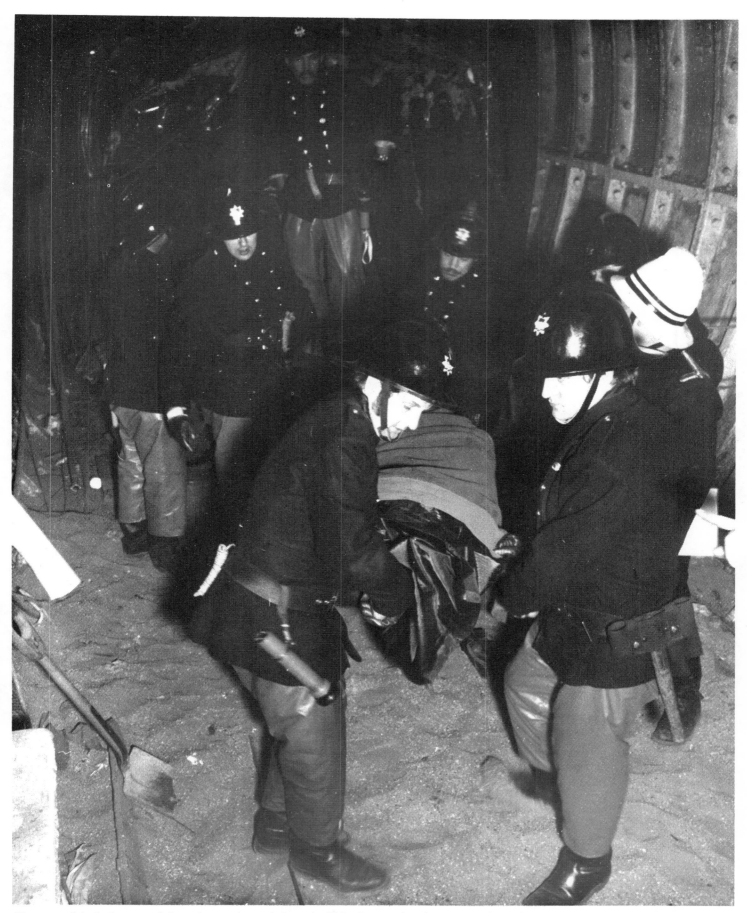

Moorgate: A body is removed from the wreckage of the train. This photograph, taken seventy feet below ground, gives some idea of the appalling conditions firemen had to work in, with little room to move, and no natural light or ventilation.
Photo: London Fire Brigade.

Bishopsgate police station, and had been obtaining emergency supplies.

At about 11.30 relief crews were sent for. Several firemen who should have finished at 9 o'clock had worked to the point of exhaustion in the sweltering heat and needed reviving themselves. Even then, they were reluctant to leave, and were back within an hour or so.

By midday 70 people had been rescued, but in the process whole sections of the upended second coach had been cut away, and as firemen were still working beneath it, special props were brought in to support it. A further 40 passengers remained to be released, including a few survivors, and as the atmosphere was deteriorating fast, an all-out effort was needed to complete the operation as quickly as possible.

After consultation with the fire brigade, London Transport staff set up fan blowers to improve the atmosphere. By 2 o'clock several more casualties had been brought out. An anaesthetist stayed in the front of the train to assist as the rescues were carried out. Only 2 known survivors now remained — a young man and a policewoman, both hopelessly trapped by the legs. It was to be a further agonizing 7 hours before the policewoman was freed, and then only by amputating her left leg. The young man was released an hour later. His ordeal had lasted 11 hours. A doctor confirmed soon afterwards that there were no other survivors.

Firemen and London Transport engineers worked together through the night using oxy-acetylene to slice through whole sections of debris which were then winched apart. By dawn the second coach was clear of casualties, but there were many bodies still to be recovered, and in the heat they had begun to decompose, making it necessary for rescuers to use antiseptic masks and gloves.

The dismantling of the second coach continued steadily all day in these unpleasant conditions, and at length London Transport ventilation experts again turned their attention to improving the air supply, and set up two large fans in the booking hall from which air was channelled down to the tunnel through plastic ducting.

Overnight efforts brought about the release of 2 more bodies early on Sunday morning. By then whole sections of the second coach had been torn away, leaving the rear of the first coach overhanging ominously, and this had to be propped up. The cutting gear was causing problems too, constantly starting fires and robbing the air of what little oxygen there was.

At this stage it was decided to fix up a kind of hoist in the tunnel ceiling to lift the wreckage clear. A battery-operated train was then shunted into the tunnel and the debris dropped into the wagon.

Conditions were now so bad that there was a danger of ready infection from the wounds the men had received during cutting operations, and they were given injections against tetanus and blood-poisoning. It was also arranged that each man should only spend 20 minutes of each hour working in the tunnel, and that there should be some form of decontamination. This was achieved by making two rough dams, one containing disinfectant and the other running water. There were separate facilities for washing the hands.

On the following day, showers were set up so that firemen and other rescuers handling the bodies could shower as soon as they arrived and change into special clothing, which could be discarded after use before dressing in their fire uniforms and returning to their stations.

Fifteen bodies now remained to be recovered, seven of them behind the fold in the front coach and five in front of it. There were two others just behind the driver's cabin, and there

was the driver himself. The props supporting the rear of the front coach were removed in the hope that it would drop and open out the fold so that the bodies could be reached, but the bogies of the first two coaches formed a wedge, preventing the coach from settling.

An attempt by the rescue team to winch the rest of the debris clear of the tunnel proved fruitless as the metalwork was tightly embedded in the tunnel wall. The rear bogie of the front coach was equally stubborn and could not be moved. All this effort had caused a deterioration in the atmosphere, prompting London Transport staff to acquire a freezer unit, which they connected to the fans in the booking hall, thus sending cool air down to the rescuers.

At 10 o'clock that night there was a bomb scare, and in the middle of their recovery efforts, firemen had to make a frustrating halt and check their appliances.

Soon afterwards the rear bogie of the first coach was pulled clear, but though the battery-operated train was brought in again in the hope of dragging more debris away, the wreckage was fused so tightly together that this had no effect.

These efforts had of course been going on all through the night, and with the end of operations in sight there was a discussion at 4.30 on Tuesday morning to decide what should be done next. For this final stage the Coroner's officials would need to be present.

When all necessary personnel had arrived, recovery recommenced, and by soon after 8 o'clock the grim task of manhandling the 7 bodies from the rear of the fold had been accomplished. After more cutting, winches were applied to the rear of the coach and the fold was opened out enough to allow access to the remaining bodies.

It took the rescue team a further 1½ hours to extricate the first five bodies that lay to the front of the fold. Another hour of painstaking effort saw the release of one more passenger, and after 2 more hours the last passenger was removed from the wreckage at 3.20 p.m.

Because details of the train's controls were needed for the pending enquiry into the accident, the process of removing the driver's body took a considerable time, and was finally accomplished shortly after 8 p.m.

The firemen's 'STOP' message was sent from the incident at twelve minutes past nine p.m. on Tuesday March 4th, over 108 hours from the time of the first message about the incident.

The Problems

Even today the name Moorgate readily calls to mind an event that claimed the lives of over 40 people in particularly horrifying circumstances. Among the numerous large-scale incidents our country has witnessed — floods, explosions, train crashes and air disasters — Moorgate remains unique for a number of reasons.

As we've seen from photographs of their rescue work, firemen are no strangers to scenes of bloodshed. Mercifully the recovery of the bodies is normally possible within a matter of hours and the incident can be brought to a close.

Moorgate was an exception. The rescue work of the first day was to give way to a further four-day period of unremitting effort to retrieve the remaining bodies whose very presence created the most appalling working conditions, deteriorating virtually from hour to hour, until finally it became necessary to decontaminate personnel.

For in occurring below ground the incident posed several major disadvantages — lack of light, space, and adequate ventilation.

The array of lighting used to illuminate the area gave off

considerable heat which, when combined with the heat from the cutting equipment and the natural exertions of the rescuers rapidly created a rank, stifling atmosphere. Even worse, conditions were so cramped that firemen had no more than about 3 feet of working space at any one time and casualties could only be removed from the remains of the coaches with the greatest difficulty, by taking the roundabout route described.

Nor were these the only problems. Equipment had to be ferried between appliances at the surface and the tunnel 70 feet below, taking up much-needed time and effort.

And finally, there was the problem of communications — vital in the matter of keeping personnel at the surface informed about progress and what further equipment was needed. The disadvantage of using normal radio below ground is that the signal is lost, resulting in indistinct messages. It was realized at an early stage in the operations that unless this problem could be overcome, the incident would become even more long-drawn out and difficult. It was decided, therefore, to bring in an experimental radio code-named 'FIGARO' undergoing development by the Home Office for work in deep locations.

The battery-operated set is worn rather than carried, as part of a special jerkin with a sewn-in aerial, and its transmitter is voice-operated, leaving the fireman's hands completely free. When used close to the metal structure of a building, the metal boosts the signal, providing clearer reception. Although several other forms of communication were used at Moorgate, this little radio proved exceptionally valuable.

Like all major incidents, Moorgate proved the need for close co-operation between the emergency services, and the support of various other organizations. Those most closely involved at the incident, included the London Fire Brigade, the London Ambulance Service, the City of London Police, London Transport personnel, medical teams from St Bartholomew's and the London Hospital, Health Authority staff, the Royal Army Ordnance Corps who manned the shower tents, and the Salvation Army who provided welcome refreshments.

Every one of these deserves full credit for the individual part they played in the rescue operations.

Yet at the end of the day it is firemen we have to thank for the fact that anyone came out of this hell alive. Their efforts at the incident, demanding the utmost from each man in selfless endurance, must surely stand as an exceptional contribution, for they were there working to free the injured and the dead, from the earliest moments to the bitter end.

Only those among us who have been similarly trapped can know what the presence of a fireman means when all hope of rescue is gone. Let me end by quoting the words of two Moorgate casualties ....

" .... I heard the swearing of the firemen on the other side of the wreckage as they battered away at it, and I called out my whereabouts to them. I knew then I would soon be in the real world outside of my tomb. Never has profanity been so reassuring."

" .... We knew as soon as we heard the firemen coming that everything would be all right, that they were going to get us out."

Such is our faith in them, and that faith is not misplaced.

CHEMICAL SPILLAGES

With industry demanding chemicals on a vast scale, the sight of chemical loads being transported along our roads is common nowadays, and it's hardly surprising if some of them come to grief. Not that all mishaps result in spillage, but when they do, they often create a serious problem.

Many of these chemicals are deadly and unless handled quickly and expertly, pose a real danger to the public. Inflammable spirit, for example, left lying in the road where people are passing by with lighted cigarettes, is an obvious recipe for disaster! So, too, are acids, which eat their way through almost anything they touch. And of course there are chemicals that kill by giving off poisonous fumes. In fact among the thousands of dangerous chemicals on file is every kind of 'nasty' you can think of.

For the fireman arriving to deal with the incident, it's important to know some facts. What kind of chemical has been spilt, and how will it behave? How should he deal with it, and what protection should he have?

To provide an instant source of basic information a special code was devised which could be shown on lorries and tankers carrying hazardous loads. Because it relates to hazardous chemicals, this code was called the HAZCHEM CODE. It uses the simple but effective idea of giving a specific meaning to certain letters of the alphabet, making the information easy to remember.

Here is the Hazchem Code as used by firemen today.

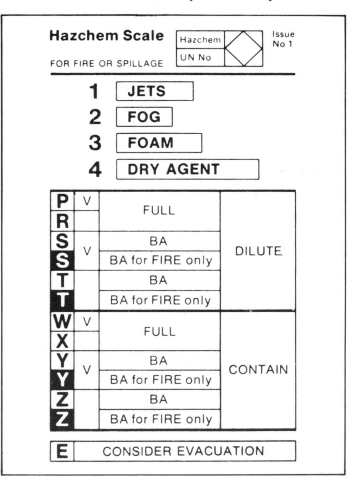

Notes for Guidance

FOG

In the absence of fog equipment a fine spray may be used.

DRY AGENT

Water **must not** be allowed to come into contact with the substance at risk.

V

Can be violently or even explosively reactive.

FULL

Full body protective clothing with BA.

BA

Breathing apparatus plus protective gloves.

DILUTE

May be washed to drain with large quantities of water.

CONTAIN

Prevent, by any means available, spillage from entering drains or water course.

Let's see how all this works in practice.

THE HAZCHEM CODE

Here is a tanker containing petroleum spirit — something we're all familiar with.

A fuel tanker displaying the 3YE Hazchem placard combining the United Nations cross-reference number, the 'flammable' warning diamond, and details of the Company for contact purposes.

Photo: Wm Cory and Son, Ltd.

At the rear of the tanker is the Hazchem label 3YE. We can easily check it against the full Hazchem Code and see what information it conveys.

The number 3 means: Use foam.

The letter Y indicates: a) Wear Breathing Apparatus
b) Prevent the spirit from entering drains.

Alongside the Y is a small 'v', showing that this is a violent chemical capable of exploding.

The letter E means: Consider evacuating the area. Whether this is done depends on individual circumstances.

Foam is used to blanket the spirit and prevent it from igniting. To stop the highly inflammable liquid from going into the drains, firemen dam it with earth or sand, and contact the company involved (whose address and telephone number are on the vehicle) so that a tanker can be sent to collect spilt fuel and any fuel left in the damaged tanker. Contaminated soil is bagged up and taken to a waste disposal site.

To assist the firemen in finding out as much as possible about a load, a PRODUCT IDENTIFICATION NUMBER appears with every Hazchem label (see photograph). Many of these identifying numbers are on file at the fireman's nearby Control Room, and an even more extensive list is kept at London Fire Brigade Headquarters. Rarer substances not on file at either of these can be traced through the computerized data bank at Harwell Research Station. Thus, by radio-ing their Control Room from an appliance at the incident, firemen can request details about any chemical from these three sources.

HOW FIREMEN DEAL WITH VARIOUS SPILT CHEMICALS

For the purpose of seeing how firemen deal with different types of chemical, I've selected a) A corrosive liquid, b) A flammable solid, c) A toxic gas, and d) A chemical requiring the use of a DRY AGENT.

a) *A Corrosive Liquid*

An example that springs readily to mind is something found in most of our homes — namely, *bleach* (SODIUM HYPO-CHLORITE).

As you probably know already, this sizzles violently on contact with certain substances, and if you get any on your skin it produces a nasty burn. It also gives off fumes that irritate the lungs. A tanker leaking thousands of gallons of this on to the road would certainly cause a nasty problem.

The Hazchem Code for it is 2P, which means that it must be damped down with a fine spray (2) to reduce the 'spitting' of the fluid, before washing it into the drains with plenty of water (DILUTE). The fireman must wear Breathing Apparatus, protective suit and gloves because of the fumes and the corrosive action of the bleach(P).

b) *A Flammable Solid*

PHENOL, our next example, is a particularly unpleasant customer. In its solid state (for there is a liquid form too), it is a white crystalline substance — a form of the cleansing agent Carbolic Acid. It is flammable, poisonous and corrosive, and gives off vapours that are very dangerous to the eyes. This chemical is unusual in that it is absorbed through the skin after contact, and attacks the body's central nervous system, causing symptoms of giddiness. Sometimes the damage proves fatal. Obviously, this is a chemical that needs handling with great care. It is coded 2X, so here again a fine spray will be used, this time to disperse the dangerous vapours. Firemen must be fully protected while they deal with it and must contain it so that it can be bagged up and taken to a waste disposal point.

c) *A Toxic Gas*

Lorries carrying cylinders of Propane or Calor Gas are a familiar sight, but numerous other gases are transported by road. One of the most dangerous is HYDROGEN CYANIDE which is used in metal plating works, and which is lethal even when inhaled in small quantities. It bears the Hazchem Code 2WE, requiring once more the use of fine spray to disperse the gas, and full protective clothing for the firemen. If you look at the Hazchem Code, you'll see that the letter 'W' is coupled with a small 'v' showing that it's a violent chemical. Because it may explode, and also because the gas itself is deadly, the area may have to be evacuated, though this will depend on circumstances. It's worth noting here that Hydrogen Cyanide is among the gases given off when polyurethane foam burns. If you have furniture in your home made of this and it is ever involved in a fire, keep well away and don't inhale the fumes.

d) *A Chemical Requiring 'Dry Agent' Treatment*

For a final example I've chosen MAGNESIUM ALUMINIUM PHOSPHIDE, (read it slowly — it's a tongue-twister!) because it's a '4WE'. This is a chemical that reacts violently with water to give off heat and poisonous gas. A serious problem would arise if it came into contact with a wet road following an accident, as this would trigger off the heat reaction and start a fire. The treatment for this substance is therefore to cover it with *dry* sand or talcum-based powder, and afterwards parcel it up in plastic sacks for disposal at a proper site. The fireman must wear full protective clothing, and because of the reaction produced by the material, will have to consider whether the area should be evacuated.

Tanks of liquid Phenol in use at a factory manufacturing plastics bearings. Note the enclosure around them to contain spilt fluid.

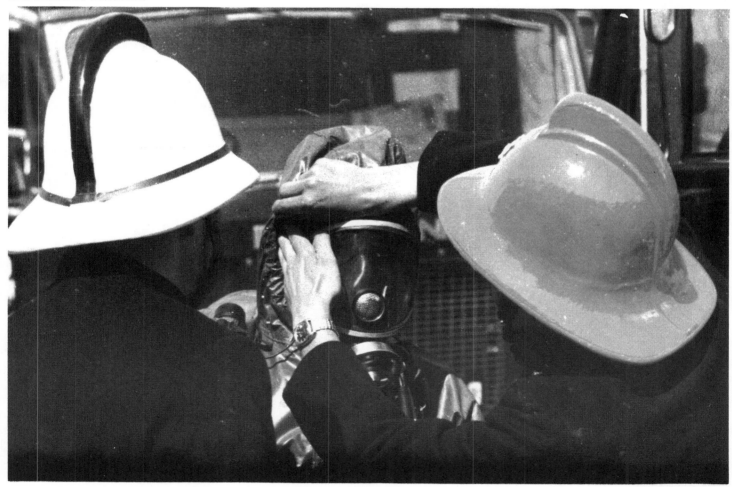

A fireman being rigged in protective clothing before dealing with a chemical spillage.   Photo: Chris Nelson.

Highly flammable Ethanol spills on to the motorway from a fractured tanker.
                                              Photo: Essex Fire Brigade.

Here, a runaway fuel tanker has rammed a parked van and overturned, spilling flammable liquid on to the road. Firemen spent a considerable time spraying water on to the spillage to prevent ignition while the cargo was transferred to another tanker. Note the placard at the rear of the tanker.
                                              Photo: Essex Fire Brigade.

A Sub Officer directs a fireman in protective clothing and Breathing Apparatus on the way he wants him to approach a chemical spill. Firemen are trained to operate fork lift trucks which are used during lengthy incidents when sometimes twenty or more drums have to be moved before the leaking container can be reached. Two more firemen can be seen preparing themselves in the background.   *Photo: Suffolk Fire Service.*

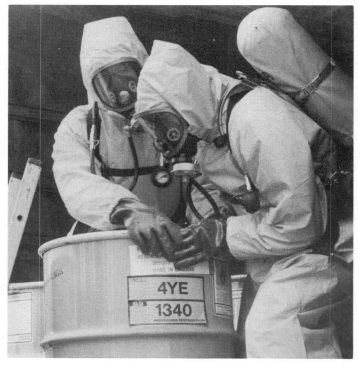

Essex firemen handling drums of Phosphorus Pentasulphide involved in an accident. This substance 'readily ignites by friction' according to the topmost label on the drums. Note the Hazchem label cross-referenced with the United Nations number, and the 'flammable' diamond on a separate label.   *Photo: Essex Fire Brigade.*

Firemen make safe a leaking drum of Hydrochloric Acid which had split in transit. A leak-sealing kit can be seen in the background.
Photo: Suffolk Fire Service.

In dockside areas accidents to cargo during handling often result in spillage. Here, Hampshire firemen are diluting a spillage of two chemicals, Chromic Acid and Phosphorus Pentachloride.

Photo: Hampshire Fire Brigade.

Firemen tackle a petrol tanker that has overturned and caught fire on a major trunk road. They can be seen here using Protein foam to control the fire. Leaking burning fuel flowed down embankments at the side of the road, presenting particular difficulties to fire crews. *Photo: Suffolk Fire Service.*

This photograph gives some idea of the lethal potential of certain chemical loads. Several families lost their homes when this tanker loaded with 5,000 gallons of petroleum spirit crashed and exploded at Westoning, Bedfordshire. Miraculously no one was seriously injured, though the driver sustained head wounds. Prompt action by firemen at spillages prevents many more scenes like this.

Photo: Bedfordshire County Council.

105

MULTI-LOADS

A multi-load is simply a load consisting of more than one chemical. In this case the Hazchem label carried will be the one indicating the greatest safety for the fireman.

To take an example, if a load were comprised of two different chemicals, one coded 2R and the other 2WE, the load would be classified as a 2WE. This means that for part of the load the precautions will be greater than are actually necessary.

UNLABELLED LOADS

From time to time firemen are called out to a spillage involving an unlabelled load. Since it would be highly dangerous to use an agent on any substance without knowing what reaction might result, the only course of action open to the fireman is to contact a chemist (e.g. ICI) and keep an eye on things until someone has been sent to identify it, which is both time-consuming and frustrating. An incident of this kind took place once in High Wycombe, when a container fell from a lorry at a busy roundabout, shattering and spilling the contents over the road. As the driver did not stop, and there was no marking on the container itself, there was no way of identifying the fluid immediately. Nevertheless it gave every indication of being something of a dangerous nature, fuming and sizzling and casting a fine mist over the whole roundabout.

Traffic was eventually halted, but not before a number of motorists had travelled through the spillage. Firemen therefore asked the police to broadcast a warning over the radio, advising those who had driven through the chemical to wash their cars thoroughly as there was a danger that the paintwork would peel off. The driver of the offending lorry was eventually traced, and in due course it transpired that the load he had aboard was concentrated battery acid — in other words, Sulphuric Acid.

Anyone walking through this would have lost not only the soles of his shoes but the soles of his feet as well!

DECONTAMINATION PROCEDURE

After attending a chemical incident, any fireman whose protective clothing has become contaminated must go through a decontamination routine in a roped-off area.

If the chemical involved is a powder, the wearer stands in a polythene sack while other firemen in protective clothing collect the powder in a vacuum cleaner. It is then bagged up for disposal. If a liquid is involved, those in contaminated suits are hosed down, again by others in protective clothing. During training this sometimes gives rise to humorous exchanges between 'victims' and onlookers, but every fireman knows that when you're dealing with deadly chemicals these safety measures are a 'must'.

Firemen wearing Breathing Apparatus and protective clothing disrobe a colleague after hosing down his contaminated chemical suit. His suit and face mask will now be sealed in plastic sacks and sent away for complete decontamination before being brought back into service.

Photo: East Sussex Fire Brigade.

THE FLIXBOROUGH DISASTER

Flixborough, a small town a few miles distant from Scunthorpe, in what was then Lincolnshire, might never have achieved any special significance, except for one fact. On a site occupying some 60 acres stood the vast chemical plant of Nypro (UK) Ltd., producing the bulk of the British nylon industry's supplies of caprolactam.

A third of this area was occupied by the plant itself and its huge stores of the numerous chemicals needed for the complex manufacturing process, among them flammable liquids in hundreds of thousands of gallons, and bulk quantities of acids and various high pressure gases: twenty acres of volatile and toxic materials.

At seven minutes to five on Saturday 1st June 1974 flammable Cyclohexane vapour ignited.

The massive explosion that followed, audible 27 miles away, demolished 90% of the plant and gave rise to fires of a savage intensity that sent up a pall of smoke a mile high. The blast brought instant death to 28 employees and injury to hundreds of others both on and off the site. Almost 2,000 buildings within a five mile radius were damaged, and at the end of the day claims arising from this catastrophe represented the largest single loss ever suffered by the British insurance market.

To anyone but those who witnessed it the scale of the havoc is unimaginable. In such a situation there is everything to be done at once: hundreds of people to be evacuated, casualties needing treatment, debris to be cleared and services to be repaired; and overshadowing all else, a raging conflagration to extinguish. In such a situation we depend first and foremost on the competence of our emergency services — our police, our ambulance and hospital personnel, and above all, on our Fire Service.

This is an account of the work of our firemen at the Flixborough incident.

The devastation following the explosion at the 'Nypro' chemical plant at Flixborough.

Photo: Press Association.

The first indications of impending disaster reached the Fire Service at 8 minutes to five, when an automatic fire call direct from the Nypro works was received at Scunthorpe Fire Station, in response to which 2 pumping appliances and an emergency tender went to the site. The explosion followed one minute later, and as firemen travelled through Scunthorpe High Street on their way to the incident they met some of the first casualties from flying glass, reporting this by radio so that ambulances could be sent. At this stage 8 further pumping appliances had been ordered to attend at the Nypro works.

Meanwhile, alerted by the sound of the explosion, the Divisional Commander of 'D' Division, Humberside Fire Brigade, had set off to take charge of the firefighting operations.

Arriving there at 6 minutes past five, he found access from the east road impossible owing to the roasting heat and the amount of debris in the road, among which were damaged high tension power cables. Blocking this approach with a private car, he gave the order for all appliances to be directed to the British Steel Corporation Wharf at the west entrance to the plant. He then sought out the Plant Manager who went with him to the west entrance, and established from him that 80 people were on the site. As further massive explosions were taking place he also determined from him what other hazards were still present, and learned that there was a natural gas main and terminal at the east entrance.

Moving from the west to the east entrance, the Divisional Commander saw that the whole plant had been seriously damaged by the blast, and casualties were escaping in all directions, many of whom received first aid from Fire Service personnel before being escorted to an assembly point at the west entrance and being taken to hospital by ambulance.

A huge fire engulfed the Cyclohexane Oxidation Plant, Caprolactam Plants 1 & 2 and storage tanks, causing intermittent explosions. At the same time the Cyclohexanone distillation towers were heavily involved in fire.

After making an assessment of the situation he sent an informative message to the Brigade Control Room: 'Major disaster, 80 people unaccounted for, numerous injured. Fire conflagration, make pumps 30, Control Unit required'. He also asked the East Midlands Gas Board and Yorkshire Electricity Board to attend.

His next task was to set up a control point for the emergency services, for which purpose he used his own radio car. All the operations were planned and directed from here.

Several courses of action were necessary at once and were carried out at the same time, but so that we can follow the progress of each one they are discussed separately.

BREATHING APPARATUS

When the Divisional Commander arrived, Breathing Apparatus teams had already been set up, each consisting of two men in breathing apparatus and protective clothing, and were searching Area 3 from a forward BA control point. Owing to the dangerous fire conditions he increased the number of men per team to five, leaving ten men in reserve, and as Breathing Apparatus operations were complex, appointed a Station Officer as officer in charge of BA operations. He also appointed a Divisional Officer as officer in charge of Search and Rescue operations, to be assisted by an Assistant Divisional Officer. Each search operation was marked on the plan of the works, and notes made of the rescue work done in both plant and buildings as it progressed. Two men were rescued by Breathing Apparatus teams.

MISSING PERSONS

Discussion with police and senior Nypro personnel established that 28 men were missing. Eight bodies were recovered from the open areas of the plant, leaving 20 who were assumed to be trapped in the plant's main Control Room, which had totally disintegrated in the initial explosion. Extrication of the bodies from this area was to prove particularly difficult as it lay beneath a vast amount of wreckage. Transformers originally situated around the outside of the Control Room had been driven into it by the blast, bringing the whole building, including heavy concrete roofs, down on top of it. On to this fell massive cableways, buried in turn by the wreckage of the plant's main pipe bridge, which included pipes of up to 61cm in diameter, all carrying various toxic, corrosive and inflammable liquids and gases.

It was decided that Fire Brigade personnel would assist the National Coal Board Mines Rescue Unit in physically locating the bodies for removal by the Coroner's Officer. The recovery work was held up for three days, however, while an unsafe Cyclohexanone distillation tower was removed, and by 19th June only 18 of the bodies had been recovered, leaving two still unaccounted for.

WATER SUPPLIES

The initial attendance officers — those first to arrive — used a water tender to draw water from an undamaged mains hydrant to supply 2 jets. Two further pumping appliances drew water from the British Steel Corporation reservoirs, their hose relaying water through more appliances near the fireground. Firemen then operated extra jets from the south-west corner of the Cyclohexane Oxidation Plant and car park. The Divisional Commander appointed a senior officer as base water officer, and as reinforcing appliances arrived each was lined up along the south boundary fence and called forward as needed.

As the reservoirs began to empty their supply was replaced by water relayed direct from 8 pumping appliances set into the River Trent. Six monitors were positioned on the south side of the plant, and one extra on the east side of Section 25A.

The river's water level began to fall with the ebb of the tide, and as a hose-laying lorry was available, a passing barge was commandeered to use as a floating pump platform. This involved shipping aboard six portable pumps from which two lines of 150 mm diameter hose were extended to the Nypro site. The first line fed two monitors cooling a 505 tonne ammonia sphere, and three jets being used to extinguish fires in the Hydrogen Plant and the Sulphate Plant and Store. The second line supplied a 25,000 litre portable dam, from which 3 pumping appliances drew water to feed 6 monitors cooling the storage tank area.

A foam attack was launched on the Cyclohexane tanks on Sunday evening, but this proved abortive as the foam broke down in the intense heat.

During the firefighting operations the Divisional Commander was approached by the Manager of the Normanby Park Steel Works, who informed him that the cooling water supplies to the steel works had failed owing to the explosion at Nypro, and that he needed 10,000 litres of water per minute within an hour to prevent the furnaces from exploding. Four major pumping appliances together with two lightweight pumps and an additional 400 lengths of hose were then sent to the steel works, where they remained with fire crews under the supervision of a Divisional Officer until the fractured water main had been repaired (Monday 3rd June 1974).

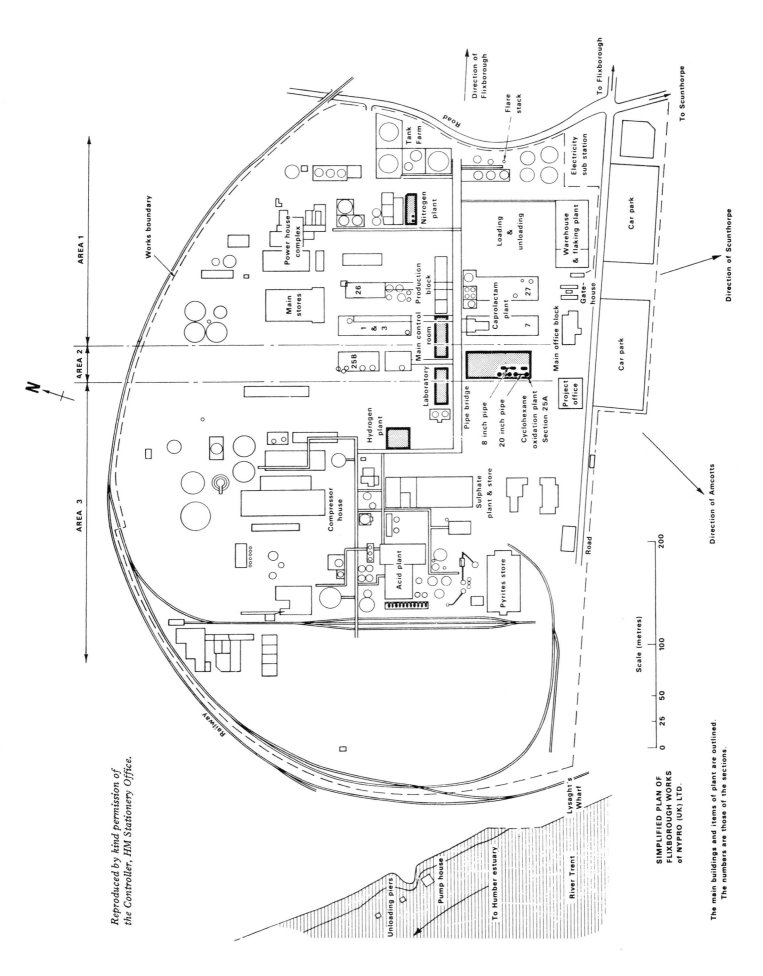

Reproduced by kind permission of the Controller, HM Stationery Office.

SIMPLIFIED PLAN OF FLIXBOROUGH WORKS of NYPRO (UK) LTD.

The main buildings and items of plant are outlined. The numbers are those of the sections.

109

FUEL FOR APPLIANCES, etc.

Extra can stocks of petrol and oil for the fire appliances were asked for in the early stages of the fire, but these proved inadequate. The Continental Oil Company were then approached, and supplied a 10,000 litre tanker of petrol and 10,000 litre tanker of diesel fuel.

Until proper meals could be organized by the British Steel Corporation the Salvation Army once again provided a valuable service in supplying refreshments for firefighting personnel.

DANGERS TO PERSONNEL

The Nypro site constituted numerous dangers to the firemen, the most obvious being the flammable materials that were present. In addition to large quantities of hydrogen, other flammables on the site at the time included over 1½ million litres of cyclohexane, 330,000 litres of naphtha, 55,000 litres of toluene, 132,000 litres of benzene, and 2,050 litres of gasoline, all giving off dangerous fumes. These were detonating continuously as firemen worked, blowing storage tanks and pipes apart so that their contents flowed uncontrolled, feeding and spreading the inferno. In fact some of these chemicals detonated without warning days after the initial fire. Fires involving tanks of these dangerous fluids were effectively extinguished by the co-ordinated efforts of firemen in dealing with each fire individually. Once the fires and leaks had been controlled, the tanks were made safe by the British Oxygen Company using inert nitrogen foam.

A number of firemen sustained burns to the skin from sulphuric and nitric acids which were leaking at different levels among the wreckage of the plant. Another hazard was a 606 tonne store of ammonia contained in above-ground spheres, which was also leaking from fractured pipes. Extra appliances and men were needed to dilute these chemicals with water sprays, the ammonia being transferred to rail tankers and removed from the site.

One of the less obvious dangers was the vast quantity of water used to extinguish the fires, which had become heavily contaminated with chemicals and turned cloudy. This lay up to 1½ metres deep in places, obscuring drains and underground storage tanks whose covers had been blown off by the blast, so that firemen were in danger of falling into them. This contaminated water also caused burns to their feet, especially those wearing leather boots, which were optional at that time. Undamaged internal effluent pumps were used to remove the water into the River Trent, under the close supervision of the River Authority and pollution officer.

And finally there was the danger from unsafe parts of the plant itself, including unstable distilling towers, and from light sheets of steel and other debris that was being blown across the site.

* * *

Reading a report of this kind gives us the opportunity of seeing what goes on behind the scenes at an incident of this size, and brings home to us the high degree of organization needed on the part of the officer in charge, and the co-ordination necessary on the part of all those under his command in order to bring the incident to its close as quickly as possible.

Only the most thorough study and regular training can equip the men of our Fire Service to deal with situations of this scale.

* * * * *

ANYTHING CAN HAPPEN . . . .

Whatever the shortcomings of other services in this country, we can boast a Fire Service that is second to none.

Our homes — every single one of them, from palace to cottage — can be traced in seconds by our fire brigade control rooms. We can put out a call asking firemen for virtually any kind of help in an emergency knowing they'll arrive minutes later, a competent, highly trained team experienced in coping with many different problems.

Just occasionally, when we read of an incident at which a fireman has been killed or injured, it comes home to us that their work is dangerous. But most of the time it escapes our notice that they pay a high price for our protection.

Most firemen have been injured in some way while carrying out their duties. Many, of course, have been killed outright or have died later from their injuries. For all their superhuman qualities firemen are, in the end, human. Their training, rigorous though it is, is no match for the real fire, terrifying and unpredictable. And when the unexpected happens, like the public they protect, firemen become victims too.

They are cut by falling slates and flying glass. Burned by hot metals, and by exploding paints and fuels. Snuffed out by sudden flashovers and electrocuted by power cables. They are crushed by heavy beams. Buried in red-hot rubble. Killed in falls. Some have been killed just getting to the fire — travelling at speed over wet or icy roads, even fire appliances can overturn. Let me show you some figures. The table on the next page is reproduced from FIRE STATISTICS U.K. 1979 by kind permission of the Home Office.

This table shows that the number of firemen injured annually has more than doubled since 1969 and now stands at around the thousand mark — a formidable figure that gives some indication of the risks involved, not only at fires, but at chemical incidents and rescues too. Moreover, this doesn't take into account the number killed on duty, or who die from diseases directly associated with their work. On average five of our firemen are killed every year. The number who die from the long-term effects of inhaling dangerous substances or repeatedly being subjected to stress will never be known.

Nevertheless we do know that smoke and toxic fumes, apart from making firemen physically sick, also leave deposits in the lungs, resulting in damage to the tissues. A considerable number of firemen die from lung diseases which may well have been caused by this damage.

Likewise, although it's impossible to say categorically that a fatal heart attack has been caused by a fireman's work, we know there's a definite link between heart disease and the stresses his work involves.

What are these stresses?

First there's physical stress. He is constantly hauling fully charged hose around and lifting heavy rescue gear, sometimes in exhausting conditions.

But more dangerous than this is heat stress arising from lack of ventilation to the body while firefighting in full uniform in a hot atmosphere. This build-up of heat causes the body temperature to rise well above normal, and if allowed to continue, leads to unconsciousness, coma, and eventually, death.

Non-fatal fire brigade casualties from fires by nature of injury 1969–79

United Kingdom   Number of persons

Year	Total–all injuries	Burns or[1] scalds	Overcome[1] by gas or smoke	Physical[1] injuries	Other	Unrecorded
1969	451	57	58	250		86
1970	446	79	72	209		86
1971	415	86	51	178		100
1972	564	121	88	228		127
1973	775	146	189	265		175
1974	658	135	87	294		142
1975	..	..	..	..	..	..
1976	1,023	162	142	373	292	54
1977[2]	974	153	91	360	157	213[3]
1978[2]	1,051	213	138	456	75	169[3]
1979	1,042	245	179	523	91	4

(1) Numbers under Burns or scalds. Overcome by gas or smoke and Physical injuries include casualties for which shock was also reported. Numbers under Physical injuries exclude casualties for which more than one type of injury other than shock was reported. Numbers under Burns or scalds and Overcome by gas or smoke for 1969 – 77 exclude casualties for which more than one type of injury other than shock was reported. Numbers under Burns or scalds and Overcome by gas or smoke from 1978 exclude casualties for which both Burns or scalds and Overcome by gas or smoke were reported but include casualties for which Physical injuries were reported jointly with Burns or scalds or Overcome by gas or smoke.

(2) Includes casualties among the armed services while firefighting during the fire service strike.

(3) Nature of injury was not recorded for most of the casualties sustained during and shortly after the fire service strike.
Not available.

There is mental stress too, which often accompanies rescue work where he's working against time to save someone's life, perhaps in harrowing circumstances. With little (if any) margin for error, there's considerable pressure on him — and remember, there's no one he can hand over to.

Finally of course, his own natural fear takes its toll. From the minute he sets out for an incident he becomes keyed up ready for action. When he arrives this tension intensifies, especially at fires where danger can come from any direction and at any time. His instinct for self-preservation sets the adrenalin pumping through his body, stimulating a supercharged alertness. After even half an hour or so of punishing effort in this unnerving environment, a fireman has spent as much energy as you and I would spend in the whole of a normal working day, and when the incident is over he's physically and mentally drained. Consider how many times this process is repeated over the years, and you begin to understand the strain it puts on his heart.

A roof collapses — and five storeys up on a turntable ladder this fireman has nowhere to go.   *Photo: Bristol United Press.*

But now we've considered the long-term effects of being a fireman, let's come back to the question of more immediate injuries and meet some of the firemen who have found themselves on the receiving end.

INJURIES

Often, while dealing with fires, firemen experience the unpleasant effects of smoke inhalation. This causes nausea and painful bouts of coughing that leave the chest feeling very tight and sore. Fumes present in the smoke sometimes cause giddiness, making it impossible for a fireman to continue working.

A.D.O. MICHAEL WHITTY

This fireman was photographed after being overcome by exhaustion and smoke inhalation during a 25-pump fire at a large warehouse. As the most senior officer present on arrival, Station Officer Michael Whitty (since temporarily promoted to Assistant Divisional Officer) had taken command, directing crews until a Senior Officer arrived to take over. Then, being the officer in charge of crews working inside the building, he joined them on the first floor, attacking the seat of the fire (its starting point). At this stage the men were not wearing Breathing Apparatus. Conditions were becoming very hot and smoky, and growing rapidly worse, until Breathing Apparatus was necessary. Men then replaced each other in crews of two after donning Breathing Apparatus. Meanwhile there was a flashover* between the first floor and upper floors, causing volumes of heat and smoke that forced the men back to the head of the stairs.

Already exhausted from firefighting, and not yet wearing Breathing Apparatus, Station Officer Whitty was overcome by the intolerable conditions and had to be escorted out. He spent an hour recovering in the ambulance and then, anxious like all firemen to get back to the job, rejoined his crews. By this time the warehouse had become too dangerous to work in as a result of the intensity of the fire, and the fire had to be tackled from the roofs of adjoining buildings.

In the treacherous conditions that prevail at fires, a fireman is open to danger from many different sources. Blinded by smoke and steam and with the roar of the fire in his ears, it is possible for him to receive an injury from an unknown source. Moreover, these same conditions may obscure the extent of his injury, as our next case shows.

ACCOUNT OF ACCIDENT TO FIREMAN PETER CULLEY DURING FIRE AT HIGH WYCOMBE FACTORY

After working on the ground floor of a burning factory, Fireman Peter Culley was about to take up a new position on the second floor when he found himself hurled to the floor and felt an acute pain in his right arm. Unable to see clearly because of the smoke and the breathing apparatus he was wearing, he picked himself up and carried on working, unaware that blood was pouring from a wound in his arm. By the time he realized something was wrong and left the building to investigate, he was on the point of losing consciousness, and after colleagues had given him First Aid, he had to be taken to hospital.

There it became apparent that something had sliced through the sleeves of his fire tunic, jumper and shirt, and had cut deep into his arm, severing nerves, tendons and muscle tissue.

*A flashover is a build-up of heat and gases that erupts suddenly like an explosion. Flashovers have cost many firemen's lives.

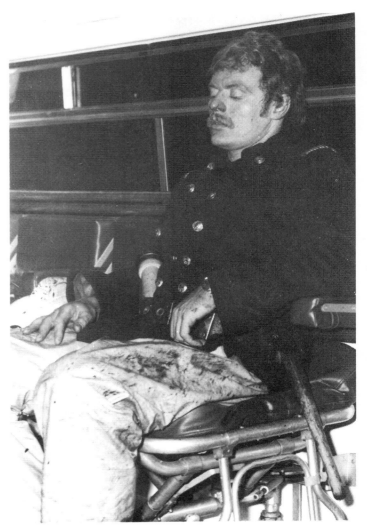

Assistant Divisional Officer Michael Whitty, photographed while recovering from smoke inhalation and exhaustion during a 25-pump warehouse fire.   *Photo: Owen Rowland.*

Fireman Peter Culley.

The injury was so severe that at one stage it was thought he might have to be invalided out of the Fire Service; however, after an operation involving micro-surgery and 3 months' intensive physiotherapy, he was gradually able to resume his former duties, though even today he experiences tingling and hypersensitivity in his injured hand.

As to the cause of the injury, this remains a mystery. One theory is that a sharp piece of heavy debris had fallen from somewhere overhead with enough force to penetrate his arm.

Another theory, which has still not been discounted, is that he was electrocuted. While carrying out an investigation later at the factory concerned, fire officers found a fire-damaged power cable hanging from the ceiling at a point near the spot where the accident happened. Closer inspection revealed that the thickness of this cable seemed to match the cut in Fireman Culley's tunic. It was also found that although power to the factory had been cut off (as is fire brigade policy), the cable was powered from a separate source, and unknown to the firemen, was still live at the time of the fire. Brushing against it in wet clothing would have been enough to result in electrocution from the powerful current.

This is an example of just one of the hidden dangers that confront firemen in the course of their work.

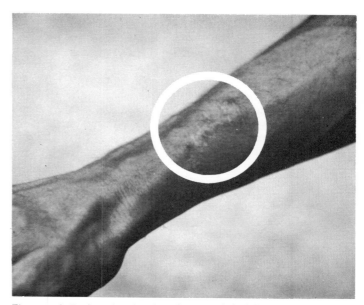

Fireman Culley's wrist showing the scars left following his accident.

There are occasions when a fireman's day begins as a very ordinary day and ends as something special.

It can be special because he saved someone's life; it can be special because he found himself dealing with an unusual job — perhaps with a lighter side to it. Or it can be special for a different reason.

It can be the day when a colleague dies at his side and he himself suffers injuries that will change his life.

Such a day came for Fireman David Whitworth in October 1978, when in the early hours of the morning a huge fire broke out at a warehouse at St Pancras Way, North London. A wall of the warehouse, weakened by the fire, collapsed suddenly as firemen were about to withdraw to a safer position. Fireman Stephen Neill, aged only 24, was killed outright, and Fireman Whitworth was so badly injured that he was not expected to live. His wife Adele gives her own account of his accident.

ACCOUNT OF ACCIDENT TO DAVID WHITWORTH DURING FIRE AT ST PANCRAS WAY, LONDON 1ST OCTOBER 1978, BY HIS WIFE

"The accident took place in the early hours of 1st October 1978. Dave had reported for work at Barbican Fire Station at 6 p.m. for his night shift. I've been told the first fire brigade machines were at the scene of the fire some time around 3 a.m., and I believe that around 5.30 – 6 o'clock a wall collapsed on the spot where some firemen were getting the machines ready so that they could be taken out of a precarious position. Dave was covered in bricks and rubble and the alarm went out that someone was hurt. Lots of firemen helped to get Dave out. For Stevie Neill it was to be death, he took the brunt of the wall, while Dave being a bit further away, escaped with multiple injuries. He was rushed to University College Hospital by ambulance.

At this time I was in the middle of feeding one of our three-month old twins at about 7 o'clock when the doorbell went. No fireman's wife needs to ask anything when a uniformed officer is standing on the doorstep. It's the moment all wives dread, they only send an officer when a fireman is dead or hurt. My only words to Mr Hale the officer were, "He's dead isn't he?"

Mr Hale was very quiet and explained that no, Dave wasn't dead, he'd merely been injured in an accident. I phoned my parents asking them to come and look after our daughters, and made a cup of tea. Funny how you reach for the kettle in a crisis.

I was taken to the hospital by Mr Hale, but when we got to Euston he was called up on the radio, and giving me the excuse that he wanted to check the exact location of the hospital we went to Euston Fire Station. I was later told he'd had a message telling him to get me to the hospital quickly as Dave was in a very bad way.

When I reached the hospital it was a mass of uniformed men. I was taken into a small room and the Senior Nursing Officer came to see me. She said the doctors would come and see me soon, and arranged for some tea.

I was allowed to see Dave for a minute. He was naked on a stainless steel bed, surrounded by nurses and doctors. He was a stranger to me, he looked so different. I was told about the suspicion of brain damage, and Dave was taken for a special brain scan. I was sent home and told to come back at 2 o'clock, which I did. What a shock! Dave was in the intensive care unit surrounded by tubes. He had tubes in his neck, nose, hands and so on, and was on a ventilator which was doing the breathing for him.

After a week he was allowed out of intensive care and put in the Neurological Ward. He was there for more than 4 weeks. He was allowed home a lot sooner than expected because I was capable of looking after him. I'd always been present and helped with his treatment.

You asked about Dave's injuries — well, he had a fractured skull, water on the brain, and a badly bruised brain. His right ear was almost severed and so badly burned that it needed two operations and many months of treatment. He was one mass of cuts and bruises. He also had burns on the insides of his elbows which were very nasty. One of his shoulders was dislocated, and he had such a bad injury to his right hand that I was told it might have to be amputated. Still, for all that, the main body organs were all intact and in good working order. We didn't know for many weeks the extent of the brain damage. But every single day that we spent at that hospital, there were firemen and officers present. I had tremendous back-up from the Brigade. They put a car at my disposal, and I was picked up each morning and driven home again late at night.

When Dave left the hospital in December he had a short break to get to know me and the house again before starting at Camden Rehabilitation Centre. He stayed there from January to July learning to use his damaged hand again. In April of '79 we had our twin daughters home again. As you can imagine that was very traumatic as there was no mother/baby or father/baby relationship between us, instead we had to rebuild that relationship from scratch.

In July Dave started back at the fire station on light duties – until October when I found him drowning in the bath. He'd had a massive epileptic fit and was unconscious. After that he was moved to 'C' Division headquarters, where he did office duties until a permanent post was found in the Breathing Apparatus workshops.

The lasting injury Dave sustained was the brain damage. It affects the part of the brain that controls memory and temper, and means that he can't remember anything that happened a short time ago. Making conversation's hard. He has a very bad temper, and the twins and I find we're constantly being abused and shouted at. Our daughters are the most insecure children you could find and naturally in the circumstances our family

life isn't entirely happy. Still, for all that we still love one another and the accident has proved how precious we are to one another. Of course, we wouldn't be able to have more children, Dave wouldn't be able to cope with any extra stress, and I have to deal with everything these days. He's supposed to get counselling from a psychiatrist.

Dave's still partly deaf, and has a few aches and pains in the shoulder. His hand is almost back to normal thanks to the wonderful work of the orthopaedic staff and physiotherapists, and the many scars are healed and fading.

Throughout all that's happened the Fire Brigade have been marvellous. Collections were made around the country and I got tremendous support from everybody. You know, we're lucky – we've got some really good friends here. Before Dave's accident, we didn't know anybody, but since then everyone's been really kind. In fact to say they've been kind is an understatement – they've been wonderful.

We've had a bit of a rough time of it I have to admit, but it's no use crying and carrying on, or wishing things were how they used to be. I try to be cheerful, but I do have a cry now and again. Mind you, I think there's worse to come, but we'll just have to make the best of it.

After all, we've still got him, haven't we?"

* * * * *

Whether the incident be a fire, a road accident or something equally serious needing their expertise, we expect firemen to get there come hell or high water. On occasions this means driving through blinding rain, fog, ice or snow as fast as can safely be managed. More than one fire appliance has come to grief in dangerous road conditions on its way to an emergency.

On 29th January 1979 at ten minutes to six in the morning, East Sussex Fire Brigade received an emergency call to a road accident on the A23 at Warninglid. Driving the Rescue Tender detailed to attend from Preston Circus Fire Station Brighton was Fireman Bob Lemon, accompanied by the senior member of his crew, Sub Officer Malcolm Walker.

Fireman Lemon had had considerable experience in driving to emergencies, but that morning there was black ice on the roads, and two miles south of Bolney, near Hickstead's famous show-jumping course, the rescue tender – in this case a Ford Transit van equipped with rescue gear – skidded on a patch of black ice. Three times Fireman Lemon managed to straighten the vehicle, but at length it slewed round and hit the verge, then ploughed through a hedge and rolled over several times before coming to rest in a field. In the process the windscreen had come out, and Sub Officer Walker was thrown out through it onto the muddy field, saved from a lethal crack on the head from a heavy fire extinguisher only by the helmet he was wearing, and which was badly damaged.

The van lay on its off side, with the driver's sliding door partly open. This had allowed Fireman Lemon's foot to slip through the gap and become trapped between the van and the frozen ground. Surprisingly, his foot suffered no serious injury. But the crash had thrown him hard against the door, jamming the handle into his spine. Still conscious, he realized he had no feeling in his legs and couldn't move them. Luckily Malcolm Walker was able to summon help over the radio, which was still working. An ambulance was despatched from Haywards Heath Ambulance Station together with a fire appliance from West Sussex. Meanwhile Bob's Station Commander, Assistant Divisional Officer Jerry Beech, had dashed to the scene and

Bob Lemon (right) with a colleague from Preston Circus Fire Station, shortly before his accident.

together with Malcolm Walker, cut a hole in the roof of the rescue tender so that he could be released. Working alongside the crew of the ambulance they took an hour to free him, and he was taken to Cuckfield Hospital where it became apparent that his spine was fractured and his spinal cord had been severed. By about 2 o'clock he was on his way by helicopter to the world-famous Spinal Unit at Stoke Mandeville, Bucks, where he was met by members of the local fire brigade.

For thirteen weeks he lay on his back, waiting for the bone to knit together. He had to be fed intravenously until the shock subsided, and was given Morphine and other powerful drugs to combat the intense pain. His bed was electrically operated, programmed to move at regular intervals to three different angles to avoid bedsores.

After thirteen weeks he was sat up, and for the first time sampled life in a wheelchair. Three months of physiotherapy followed to revitalize his wasted muscles.

It's unlikely that Bob Lemon will ever walk again. Today he is paralysed from the waist down, and will always be under Stoke Mandeville's care. By sheer perseverance he has trained himself to get around by making a kind of hopping movement

Bob gets a visit from his wife and family in Stoke Mandeville Hospital. Paralysed from the waist down, he now spends most of his life in a wheelchair.

Photo: Brighton Evening Argus.

115

with the aid of his calipers, just about the house and garden. For longer journeys he uses a car with special hand controls which the Fire Services National Benevolent Fund helped him to buy. This enables him to take his family out and gives him a measure of freedom away from his wheelchair.

Bob has recently taken up photography, and has already won two major prizes. Friends helped him build a darkroom in his garden and he has since equipped it with some of the money awarded him as compensation for his accident.

Bob is full of praise for the help he received, both from the fire brigade and the public following his injury. Thousands of pounds were raised for him by well-wishers, among them Jimmy Saville, to whom Bob later made a special presentation.

Firemen from Aylesbury and other Bucks fire stations visited him at Stoke Mandeville. Among others who visited him there was one of the ambulancemen who first helped to rescue him. Friends and relatives too, helped in any way they could.

The courage and cheerfulness of Bob and his wife are remarkable — all the more so since the youngest of their five children, six-year-old Colette, has had more than twenty operations for a rare internal disorder. For the Lemon family, hospitals seem to have become part of everyday life.

Today Bob Lemon is no longer a fireman. Yet that same courage and strength of character that made him become one shines through loud and clear.

Two firemen died in this fire in Reading while working on the ground floor, which collapsed into the basement.   Photo: Royal Berkshire Fire Brigade.

These casualties represent only a few of the risks our firemen run from day to day. During the years 1971-1980 alone we have lost 38 firemen, either killed on duty or fatally injured. Several more have already died since. The cruel fact is, there will be others — perhaps even from your local fire station or mine. They all take the same risks.

How many of us would be willing to exchange our uneventful way of life for the uncertainties of theirs?

THE FIRE SERVICES NATIONAL BENEVOLENT FUND

The Fire Services National Benevolent Fund, first launched in 1943, exists to help any member of the Fire Service in genuine need, especially firemen in need of financial help through injury, and dependants of those who have died.

This remarkable fund derives its income solely from voluntary contributions both from members of the Fire Service and the general public and puts this money to a wide variety of uses.

For example, immediate money grants are made to the families of those killed or injured so that their financial difficulties are eased at the crucial time.

Money is also provided to buy specially adapted motor cars for personnel totally disabled through Service injuries.

A special fund set up for widows and orphans makes grants towards the maintenance of these children throughout their education.

And for firemen recovering from injury or ill-health there is a convalescent home at Littlehampton, Sussex, also maintained by the Benevolent Fund. The home, known as Marine Court,

consists of 24 fully furnished flats, where personnel can stay free of charge with their wives and families, providing only their own food. An extension to this, Munson House, accommodates six firemen travelling unaccompanied.

More recently, Harcombe House at Chudleigh in Devon has also been acquired by the Benevolent Fund as a convalescent home. Staff at all these make firemen and their families welcome during their stay and see that all their needs are taken care of.

An outsider to the Fire Service myself, I have both heard and seen statements of gratitude from injured men and bereaved families testifying to the wonderful help they have received from the Benevolent Fund, and the real concern shown by those who run it.

Firemen, who so frequently involve themselves in money-raising ventures for other charities, also hold Open Days and other events with their own fund in mind.

It is a cause that deserves wholehearted support.

Marine Court, the convalescent rest home for firemen at Littlehampton, Sussex.

FIRE BRIGADE CONTROL ROOMS

You may find it hard to believe, but most of us at some time will call on the Fire Service for assistance. One in every four of us can expect to be involved in a fire, and as you've already seen from earlier chapters, this is only one of the accidents that can happen to us.

Of course, you may be one of the lucky ones. Even so, you might need to call the fire brigade for someone else, and then it will be up to you to give them all the information they ask for, so that they can send help as soon as possible. All of which sounds easy! But fire sometimes makes people panic, and then they do all kinds of odd things — like forgetting to give the address, or rushing their words so that the message is misunderstood. So when you make your 999 call, rule number one is: Keep your head.

It's as well to understand that when the operator connects you with the fire brigade, you won't, as you might expect, be speaking to your local fire station. In fact you'll be speaking to an operator in a fire brigade Control Room, which may be some distance from where you live. This Control Room will accept emergency calls over a wide area, and will be responsible for sending out fire appliances from perhaps as many as thirty or forty different fire stations.

Sometimes a caller, thinking he's speaking to his local fire station, tells the operator that the fire is 'just around the corner' and promptly rings off without giving any details at all! Valuable minutes are lost while the Control Operator stands by in the hope that someone else will make a call and give him the information he needs. As you can imagine, a delay like this

A fire brigade control room where information is retrieved manually, showing desk units at which control officers take incoming calls. Coloured lights on the map of the area show appliances available at various fire stations.

could mean the difference between saving a life and losing it. In relation to the number of emergency calls they handle, Control Rooms are quite small (perhaps twice or three times the size of the average family sitting-room) and are manned by just a handful of control staff.

In the centre of the room are several desk units like switchboards at which the control staff sit while they take the calls.

On the wall in full view of these is a large map of the area, and on it, just below the names of various towns, are lights of different colour, each one representing a particular kind of appliance available at the fire station there. Some fire stations have as many as six machines (6 lights) and others just one (1 light), depending on the risks the fire station has to cover. When an appliance goes out to an emergency its light is switched off,

A control officer consults a predetermined Attendance Card showing which fire station will deal with the incident and which fire appliances will attend.

By pushing the 'turn out' button the control officer mobilizes the fire station concerned and gives particulars of the incident.

A control room where MICROFICHE READERS are in use — these are the two 'television screens' you can see. As soon as the operator receives the address of the incident, part of this address is keyed into the machine, which rapidly displays the predetermined attendance, next nearest fire stations and fire appliances, etc.
Photo: Tyne & Wear Metropolitan Fire Brigade.

showing that it is committed.

Equipment in some Control Rooms is more sophisticated than in others, so the method of dealing with calls differs slightly. First, a control room where information is retrieved by hand.

Imagine now that you've made your 999 call and the operator has put you through to the fire brigade. In the Control Room a buzzer alerts the Control staff and an operator takes down your information on a note-pad, to be transferred later to a special form. Meanwhile, from a card index system a Control Officer quickly selects the card with the name of your street on it. This card gives certain information: the fire station nearest your home, the shortest route there and what appliances must attend. This is called a PREDETERMINED ATTENDANCE. But wait — there's more! The card also lists the six next nearest fire stations in case extra help is required, and where special appliances such as a Turntable Ladder or Emergency Tender can be found. Nothing is left to chance.

With this information to hand, the Control Officer alerts your local fire station by direct telephone line and passes on the particulars.

At the fire station the fire bell sounds. Coloured lights come on showing which appliances are due out, and whatever the firemen happen to be doing — whether they're at a lecture, doing a drill or having a meal — they'll drop everything and be on their way to you in seconds.

While they're dealing with the incident, firemen keep radio contact with the Control Room from their appliances, and can ask for extra man-power if necessary. When the fire is under control they'll send a 'STOP' message.

Some fire brigades have installed Visual Display Units in their Control Rooms. These have switchboard units like the ones I've already described, each with its own electric typewriter and a monitor screen above, like a small TV set. In this case when you make your call, the Control Operator types out your information, which then appears on the screen above and others next to it, so that the Control Officer can see the details at a glance and look out the Predetermined Attendance card while the operator is typing your name and telephone number. Once the card has been traced, the operator types out the instructions from it and transmits the complete message to teleprinters at your local fire station, automatically setting off the fire bell. Amazingly enough it will have taken only about 20 seconds from the time you made your call!

Even more sophisticated than this are Control Rooms using MICROFICHE READERS. These machines also have screens like a television set, but work differently from Visual Display Units. Here, as soon as the operator receives the address of an incident, part of the address is keyed into the machine, which then rapidly displays details of the Predetermined Attendance, next nearest fire stations and appliances, etc. This information is stored on minute negatives each only about a quarter of an inch long, set out in batches of perhaps 30 or so on postcard-size negatives (fiches) in a carousel type container.

But sophisticated though these systems are, we live in the age of the computer, and in common with many organizations the Fire Service will eventually be replacing much of their equipment with computers, as the photographs overleaf show.

To go back to the Visual Display Units, I was lucky enough to watch these at work in a modern Control Room. It was fascinating, but at the same time sobering, to read the details of various emergencies on the tiny screens: a large fire at a hairdresser's in Beckenham . . . . a lorry in collision with petrol

pumps at a filling station . . . . an overturned tanker leaking chemicals . . . . a two-vehicle collision, three persons trapped . . . . an overturned lorry, one person trapped under load . . . . a nine-year-old boy locked out of home . . . .

So the list continued. All those people were depending on the firemen. But they were depending too on the staff in that Control Room — the vital link between people in trouble and the men who would help them. Like the firemen, Control staff cover us round the clock, cool-headedly persuading distressed callers to give them clear information so that help can be sent at once. I can't emphasize enough how much we depend on them. No praise is too high for them and the work they do behind the scenes.

FIRE PREVENTION WORK

In the same way that there are provisions for our safety on the roads, so there are provisions for our safety in public buildings and places of work.

Take a look at these illustrations

We pass safety points like these almost daily, probably not giving a thought to how or why they came to be there. Seldom do we think about the people who make sure these safety measures exist and are kept in working order, yet the work they do is essential to us all. Like the firemen they save lives, but in a different way.

Who are they? The answer is, our Fire Prevention team — Firemen attached to a special department with the responsibility for seeing that the Fire Safety Laws drawn up in Parliament are obeyed.

Fire Prevention work falls broadly into two categories. First, there is the enforcement of fire laws relating to specific buildings, which means ensuring that people can escape from them safely if a fire occurs. These include offices, shops, factories, hotels and boarding houses, and also railway premises.

Secondly, there's what we might call 'goodwill' advice. That's to say, the inspection of premises not covered by these laws but which come under the legislation of other authorities and have to maintain certain standards required by them. This includes such buildings as schools and hospitals, places of entertainment, churches and private houses.

To take some examples, places of entertainment normally come under the jurisdiction of the District Councils, and the Fire Prevention Department inspects these premises for them. Buildings are also governed by special Building Regulations, and Fire Prevention Officers may be called in for advice — for example, as to whether in exceptional circumstances a safety measure incorporated into the plan of a given building may be waived providing that a satisfactory alternative arrangement is made.

Insurance companies too have an interest in fire safety — they like to know that a building is reasonably well protected before they insure it, and will go so far as to offer reduced premiums to property owners who instal extra precautions such as sprinkler systems.

The fact that numerous authorities are involved means that fire prevention work is very complex.

Let's see what the rules and regulations amount to, and how they touch on our everyday lives.

It is important that all buildings are made as safe as possible from fire, and that wherever practical fire-resisting materials are used in their construction. In public buildings this is even more important. Devices for giving early warning of fire, escape routes, fire doors to halt the spread of smoke and flames, and firefighting aids like hose reels and extinguishers are all essential where people gather in large numbers, and Fire Prevention Officers must make regular checks to see that these safety measures are being observed.

Take an office block for instance. Here they must check that there are enough exits to allow the safe escape of the people that work there — and not only that there are *enough* exits, but also that they are a certain width, clearly marked and well lit, and free of obstruction. They also check that fire safety notices are on view, explaining what should be done in the event of fire.

You have all been involved at some time in a fire drill, either as a pupil at school, factory employee, or member of staff at a department store or hospital. These drills must be held regularly and a record of them be kept in a special log book, which is checked by Fire Prevention Officers when they visit.

Moving on to the Building Regulations I mentioned earlier, these require that buildings be *compartmented* — in other words, divided into smaller areas by the use of walls and partitions to reduce fire spread. Where houses are terraced, there must be a solid 'PARTY WALL' between properties,

Fire safety is their business . . . . This photograph shows just how much concern is given to our safety by our Fire Prevention departments. On the far side of the room above shelves filled with rolled-up building plans, a plan is under discussion. On the desks and in the filing cabinets are books and documents on fire legislation, advisory leaflets and so on. Note the chart of Hazard Warning Diamonds on the far wall.

again to contain fire within a limited area.

Fire Prevention Officers carry out the work of checking all these precautions, both on premises required by law to hold a Fire Certificate, and on premises where a particular authority will not grant a licence unless fire precautions are satisfactory. When you consider how many buildings come within one category or the other, you realize what a monumental task they have.

Other responsibilities include acting as the Petroleum Licensing Authority for county councils, which involves the inspection of petroleum installations already in existence, as well as others being built.

A senior Fire Prevention Officer must attend any fatal fire and any petroleum spillage in his area.

In connection with the Fire Precautions Act, a lot of the Fire Prevention Officer's work consists of discussing buildings with architects while they are still in the planning stage, so that they can recommend safety features that should be incorporated as the building gets under way. It can then be inspected by them at various stages of development, and then the granting of a Fire Certificate presents no problem.

Ideally this would happen with all new buildings, and ideally once they had materialized they would remain unchanged. Unfortunately the true picture is not so rosy! It's not unusual for architects to submit plans that fail to conform to the basic fire safety regulations, and then it falls to the Fire Prevention Officers to mark satisfactory alternatives on them

using a red pen.

And of course buildings change hands from time to time and are altered to serve different purposes. Some are eventually divided up into separate premises, each requiring individual fire precautions. From the regular checks they make, the Fire Prevention Officers note any change in construction or usage that may affect the safety of people in those buildings. They follow up by seeking out the relevant plan from their files and mark on it any extra safety measures the owner must provide to meet the requirements of the law.

This means, as you might expect, that Fire Prevention Officers aren't exactly welcomed with open arms by the owners of the premises they visit! It's a nuisance having them cause a disturbance when you're trying to run a factory, and the alterations they want you to make will mean a lot of mess and inconvenience. Anyway, who wants to spend all that money for nothing? – we couldn't possibly have a fire *here!*

Yes, we all seem to think that. But fire does happen to someone, somewhere, every day. When it happens in buildings such as I have described and the fire precautions are not up to standard, there is often loss of life. Time and time again we read of bodies piled up at padlocked fire exits, just a few steps from fresh air and safety; of people trampled to death at *open* doors by others who panicked as they found other fire exits blocked. And how many times have we heard of people trapped in upstairs rooms while the one and only escape stairway burned? Many of these deaths were preventable, but the fire safety regulations were ignored.

However, flouting the rules brings heavy penalties. Property owners who refuse to toe the line when it comes to maintaining adequate fire precautions on their premises can be heavily fined and sent to prison — especially if people have died as a result of their negligence. For if, on a further visit, the Fire Prevention Officers find that their requirements for a given building haven't been carried out, they report this to the Chief Fire Officer, who in turn notifies the County Council and prosecution follows.

But fines and imprisonment don't bring victims back to life, and Fire Prevention Officers can be forgiven for feeling bitter when people die because their advice has been ignored. As firemen, they've seen the horrifying results of fire, and they care enough about our safety to put all their energies into teaching us how to prevent it.

When they give us advice, the least we can do is *listen*. And here is some of that advice:

A deadly combination — an overloaded socket and frayed flex. Countless fires have resulted from simple faults like these.

Doing a spot of decorating? If you smoke near flammable thinners you may end up with no house to decorate!

This chip pan was too full — and why was no one keeping an eye on it?

It was fun lighting matches, until they burned their fingers and dropped one.

She's been lucky before. This time? An unseen toy, a slip, and burning fuel will flow everywhere.

Full marks here! This fire is protected by a guard and is free of clutter that could scorch and catch fire.

Fires and casualties from fires in dwellings by source of ignition

United Kingdom 1979	Fires	Number	
		Casualties	
Source of ignition		Non-fatal	Fatal
Total	58,640	6,075	865
Smokers' materials, matches	5,802	1,252	265
Space heating	6,559	1,054	199
Electric	1,997	338	89
Solid Fuel	1,733	256	53
Gas (mains)	1,146	219	28
Oil and petroleum	958	126	20
Other	725	115	9
Malicious or doubtful	3,234	483	62
Cooking appliances	20,382	1,470	47
Electric	14,357	914	24
Gas (mains)	5,409	515	22
Other	616	41	1
Electric blanket	1,926	183	44
Children with fire	4,261	359	29
Electrical wiring	3,921	203	24
Naked light, taper, candle etc.	682	129	23
Other	8,965	458	43
Unknown	2,908	484	129

The room on the other side of the doors shown in the photo below, remained undamaged apart from a small amount of smoke damage above the doors. This is why it is always wise to keep doors closed.

Possible cause of this fire — an overturned lighted candle being used to give light after the electricity supply had been cut off. Note that damage is worst in the upper half of the room, proving that hot air and gases rise. The closed doors on the right are important (see photograph above).

The heat from this kitchen fire has stripped tiles from the wall by the cooker and broken the windows. Note how fire has penetrated cupboards, bursting cans inside, several of which can be seen lying on the floor.
Photo: Tyne & Wear Metropolitan Fire Brigade.

FIRE SAFETY SUGGESTIONS

CHILDREN are fascinated by fire — and often killed by it. Playing with matches or poking things into a fire is dangerous, and is one of the main causes of fire in this country. Children should always be closely supervised where any type of fire is in use, bonfires included.

SMOKING can endanger your health in more ways than one! You may still be enjoying a cigarette when you're a hundred. Unless you drop off to sleep while you're smoking in bed or get careless with those lighted cigarette ends. Then you could start a fire and die sooner than you think. DON'T smoke in bed. MAKE SURE cigarette ends are safely stubbed out in an ashtray.

ELECTRICITY is an unseen friend — or enemy. It all depends on how wisely you use it. Don't abuse it by:
a) Overloading sockets with several plugs.
b) Using frayed flexes.
c) Tampering with electrical wiring.
d) Misusing any electrical appliance. Makers provide instructions, and they do know best!

HEATING: This needs special care as people are always moving around close by.
a) Open fires need a fire guard to keep sparks in and danger sources out — and this includes people. Chimney fires cause damage to a house and may lead to a more serious fire, so have chimneys swept at least once a year, and more often if the fire is in daily use.

b) PARAFFIN fires need a safe place where they won't be knocked over — preferably fixed to a wall.
DON'T FILL THEM while they're alight.
DON'T MOVE THEM while they're alight.
c) Electric fires must stand well back from anything that will burn. The lead should be tidy, as near to the wall as possible, so that people don't trip over it and fall onto the fire.
d) Gas fires too must stand well away from anything that can catch fire. Follow the instructions when changing the cylinder. Fixed gas fires need care when lighting. Make sure you're ready to light the fire as soon as the gas is turned on.

COOKING FAT can overheat or overflow, so keep an eye on anything being cooked in fat. When putting food into hot fat, it's a good idea to remove the pan from the heat and stand it on a safe, level surface while you do it.

FLAMMABLE SUBSTANCES can be found in most homes — polishes in aerosol form, paint thinners, paraffin and so on. Keep them to a minimum, and see that they're not placed near anything that can ignite them.

RUBBISH is a ready source of fuel if fire breaks out. It tends to accumulate around the house, so keep it tidied up — better still, clear it out altogether.

FIRE SAFETY IS JUST GOOD COMMON SENSE. Remember! Someone, somewhere will have a fire on their hands today. DON'T LET IT BE *YOU.*

Children had been left unsupervised in this house.

Photo: Walsall Observer.

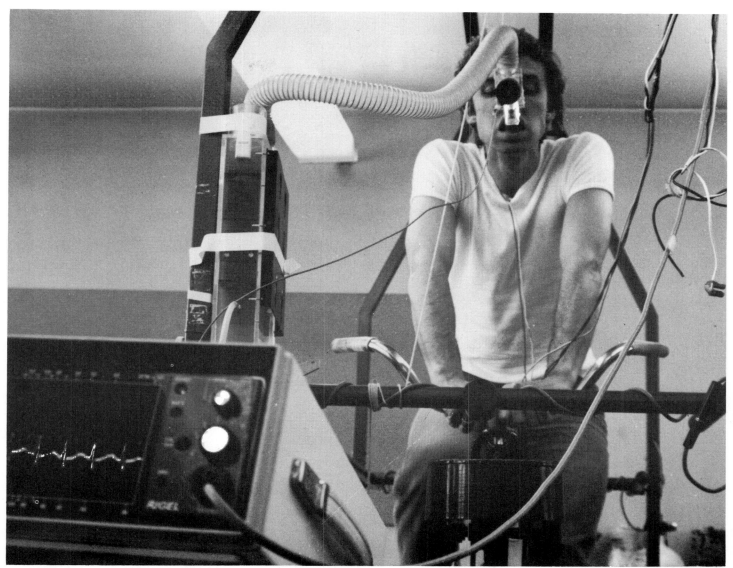

Physical fitness is essential to the fireman. Here a member of Tyne & Wear Fire Brigade is using a physiology and respiratory unit during tests aimed at finding the best way of achieving and maintaining fitness.   *Photo: Tyne & Wear Metropolitan Fire Brigade.*

SO YOU WANT TO BE A FIREMAN?

In case there are any budding firemen among you, here are a few facts you may like to know before you apply for the job. To begin with, you need to be between 18 and 30 years old — unless you've done military service, in which case you're eligible up to the age of 34. Next, you need to be 5ft 6ins tall or over. Your chest must measure at least 36ins and be able to expand 2ins — this is most important.

If you measure up so far, the next hurdle is a test involving Arithmetic, English, Dictation and physical fitness, and if you pass this, you'll be sent off to do 3 months' basic training. During that time you'll have to learn to wear Breathing Apparatus. This is an absolute 'must' — no one can become a fireman without it. If all goes well you'll spend the next 15 months on probation learning all about the job and taking a written exam every three months. With luck you'll get some experience at incidents too, while the rest of the watch take you under their wing. After 2 years as an operational fireman, and having passed the necessary exams — written and practical this time — you're eligible for promotion to Leading Fireman. To

become a Sub Officer you'll need to do another 2 years' service and pass more written and practical exams. A further year's operational service with a qualifying written exam makes you eligible for promotion to Station Officer.

As you move up the ladder you'll take part in various technical and practical courses in Firemanship at the Fire Service College at Moreton-in-Marsh, Gloucestershire. The College accommodates up to 500 residential students, and in its extensive grounds there are training facilities of every kind — industrial and domestic set-ups, a Breathing Apparatus Training block, a transformer house — even an imitation ship complete with 'sea'! The training buildings are built of fire-resisting materials so that full-scale fires can be set in them over and over again without damaging the structure.

The 'fireground' at the college is reckoned to be the most advanced anywhere in the world, and students of every nationality come here for their training. This alone is proof that the training of our firemen is something we have every reason to be proud of.

Students of the Fire Service College leaving classes in the administration building to go to lunch.

Photo: Eric George.

128

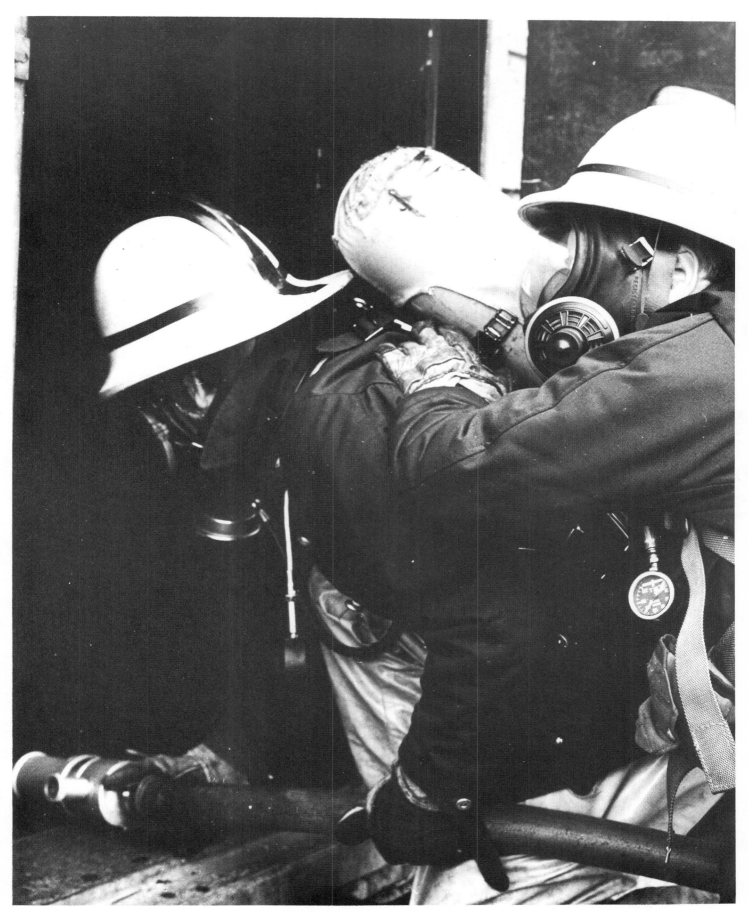

Breathing apparatus training in progress — men about to enter one of the College's training buildings, constructed of fireproof materials.

Photo: Eric George.

An overseas student takes part in a fireground exercise.

Photo: Eric George.

An exercise on the simulated ship, which is built of concrete. Alongside this, but out of sight, is a stretch of water acting as the sea.   *Photo: Eric George.*

The fireground at the College also has its own oil tanks to provide training facilities. Here, students are being instructed in the correct use of the foam monitor.

Photo: Eric George.

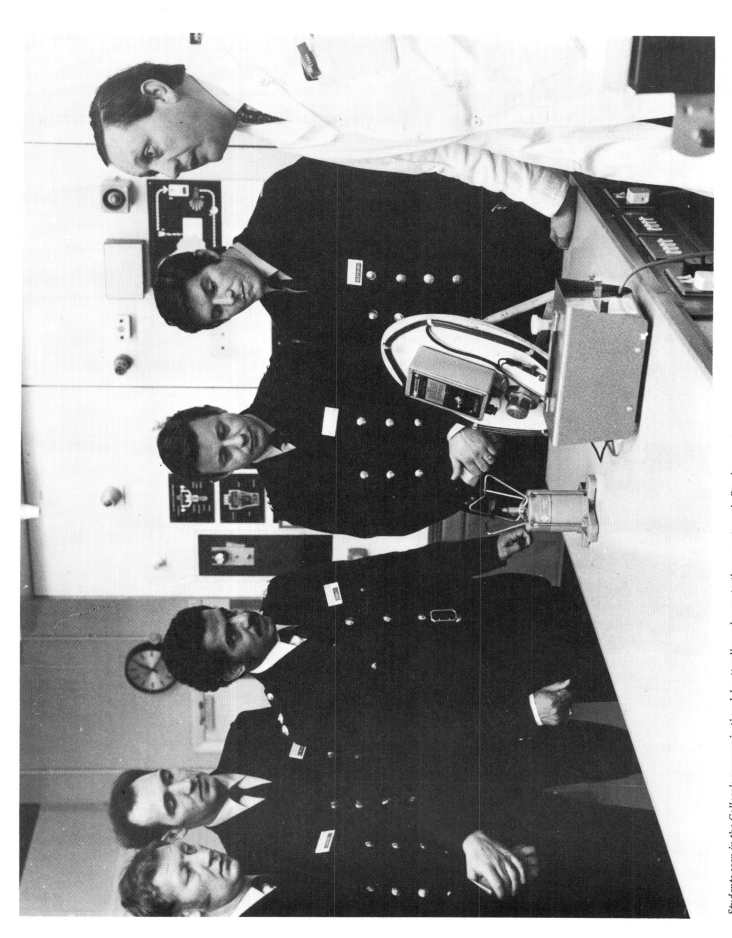

Students seen in the College's communications lab. attending a demonstration on automatic fire alarm systems.

133

FOOD FOR THOUGHT

The sheer importance and urgency of their work precludes us from going behind the scenes to find out for ourselves what firemen do, and we outsiders can perhaps be forgiven for having underestimated them, for having misjudged them. There are many, for instance, who imagine that firemen spend their day killing time while they wait for the next incident to happen. I hope we can do something to dispel this myth.

Firemen, too, are their own worst enemies — their reluctance to talk about their achievements or to seek thanks for themselves, works against them, and their work at incidents is frequently misreported by the Press, or simply not reported at all. On this subject, to anyone responsible for reporting facts to the public about fires or other incidents where firemen are involved, I would offer this advice: Bide your time until you can see that the incident is under control, then ask to have a word with the officer in charge. He will have all the facts — what casualties there were, how much damage was caused, the number of personnel and appliances in attendance, and so on. Firemen have put up with inaccurate reports long enough. They'll thank you for it if you get your facts right.

* * * * *

To me firemen have always been a symbol of all that's best: brave men whose concern is the safety of others, who disregard their own discomforts to extend a compassionate helping hand to anyone who needs it. Men in whom we can have complete faith.

I believed this as a child, and I still believe it. Having drawn closer to them, having watched them at work from within the family circle as it were, I know how true it is.

I can never pass a fire station — much less enter one — without a feeling of awe, an awareness of all that we owe to the men inside. These are men whose lives are so different from ours. They live with danger and they look on death and disfigurement a thousand times over. I find myself thinking of all the people who are alive today because of the work they do.

If you hear the two-tone horns today and you see them go by, I hope in your imagination you'll go with them; share their apprehension as they climb into their fire gear and head for their destination. Will it be some huge warehouse fire? Another tragedy on the motorway? A road tanker leaking deadly acid? Or some frightened child trapped by his own mischief?

I hope you'll picture them later as they return to their fire stations, wearily discarding their wet clothing. Going home to their families, some with a cheerful greeting; some too distressed by the day's incidents to speak or eat.

I hope you'll think of the ones who got injured today and won't be going home; and of the ones who lost their lives, some of them still very young, because they wanted to be of service to you and me.

No one, I repeat, no one, makes a more valuable contribution to this country than our firemen. We should be proud of every one of them.

Photo: Tony Cairns.